Mathcad
A Tool for Engineering Problem Solving

Second Edition

Philip J. Pritchard
Manhattan College

McGraw-Hill
Higher Education

Boston   Burr Ridge, IL   Dubuque, IA   New York   San Francisco   St. Louis
Bangkok   Bogotá   Caracas   Kuala Lumpur   Lisbon   London   Madrid   Mexico City
Milan   Montreal   New Delhi   Santiago   Seoul   Singapore   Sydney   Taipei   Toronto

McGraw-Hill
Higher Education

MATHCAD, SECOND EDITION (BEST SERIES): A TOOL FOR ENGINEERING PROBLEM SOLVING

1 2 3 4 5 6 7 8 9 0 DOC/DOC 0 9 8 7

ISBN 978–0–07–319185–0
MHID 0–07–319185–X

Global Publisher: *Raghothaman Srinivasan*
Executive Editor: *Michael Hackett*
Senior Sponsoring Editor: *Bill Stenquist*
Director of Development: *Kristine Tibbetts*
Outside Developmental Services: *Lachina Publishing Services*
Executive Marketing Manager: *Michael Weitz*
Senior Project Manager: *Kay J. Brimeyer*
Senior Production Supervisor: *Kara Kudronowicz*
Associate Media Producer: *Christina Nelson*
Associate Design Coordinator: *Brenda A. Rolwes*
Cover Design*: Studio Montage, St. Louis, Missouri*
(USE) Cover Image: *Royalty-Free/Getty Images*
Compositor: *Lachina Publishing Services*
Typeface: *Century Schoolbook 10/12*
Printer: *R. R. Donnelley Crawfordsville, IN*

MathCAD is a registered TM of the PTC Corporation.

Library of Congress Cataloging-in-Publication Data

Pritchard, Philip J.
  Mathcad : a tool for engineering problem solving / Philip J. Pritchard. -- 2nd ed.
    p. cm.
  Includes index.
  ISBN 978-0-07-319185-0 --- ISBN 0-07-319185-X (hard copy : alk. paper)
    1. Engineering--Data processing. 2. MathCAD. I. Title.
TA345.P765 2008
620.00285'536--dc22
                                          2006035839

www.mhhe.com

Contents

Appendix: Graphing 171

Preface

This second edition was written as part of McGraw-Hill's *BEST* series (Basic Engineering Series and Tools). The intended audience for all of the books in this series is the introductory engineering class. However, as I was writing the book, it became apparent that, due to the nature of Mathcad itself, even a text covering only the main features would end up covering some material to which a beginning engineering student might not yet have been introduced. Hence, although the primary market for this text is still the introductory engineering class, in truth engineering students from freshmen to seniors, and even graduate students, will find it helpful in learning Mathcad. Practicing engineers who want a compact guide to using Mathcad for solving their engineering problems will probably find it useful as well.

Many colleges in the United States are now using a one- or two-semester course designed to introduce students to the basics of what it means to be an engineer. These courses invariably include exposing the students to use of the personal computer as a communication and analysis tool. For example, they may be taught how to use a spreadsheet, a technical calculation package such as Mathcad, and a programming language such as Visual Basic for doing engineering work. This book is suitable for use in such a course.

Each chapter (with the exception of the first) introduces features of Mathcad by demonstrating a variety of very practical engineering examples, so that the reader can see that the features being described do have a real-world, practical engineering application. The best approach is to read each chapter while at the computer, so that the reader can replicate each of the examples. In doing so, they will not only learn how to use Mathcad (and, perhaps more importantly, how *not* to use Mathcad), but also get exposure to some typical engineering problem-solving approaches and methodologies.

Chapters 2–7 have sets of exercises. These are intended as practice exercises on the specific material covered in a section or chapter; some exercises introduce topics not specifically covered in the chapter. It's probably a good idea that the reader do all of the exercises. Most of these would also make good homework questions.

Freshmen, sophomores, juniors, and seniors will each find the book useful in learning Mathcad. Regardless of any previous experience with Mathcad, readers should be able to move fairly quickly through much of the material and become increasingly comfortable with Mathcad. The book is intended for people who have never used Mathcad, as well as those who have some knowledge of it. The former will be able to self-teach by reading the book from the beginning. The latter will be able to pick and choose those topics they wish to learn more about. As always, each new edition of Mathcad has a number of important changes from previous versions, not the least of which are the computing engines behind the math, equation-editing techniques, and ways of manipulating math regions; hence, even experienced Mathcad users will probably receive some benefit from this book, at least as a reference tool.

Graduate students and engineers in industry should have no difficulty in reading this book from cover to cover (while using the computer) in a relatively brief period of time. After doing so, they will have a good grounding in the basic features of Mathcad and a sense of its power in solving the more advanced engineering problems that they will probably be involved with.

What is Mathcad, anyway? We'll discuss this in detail in the first few chapters, but here let's just say it's one of several very powerful technical calculation applications that are available today. Although it is always difficult to make such predictions, it seems that Mathcad will continue to be a dominant application of this kind. In my opinion, if it does so, it will be well deserved. Mathcad started out back in the days of the non-graphic DOS computer world. You may have experienced technical calculation applications that are very "clunky," that have some odd or crude interface features such as using a command line, or that have a very odd menu structure or icon system (I don't want to name names, but if you've used the ones I have in mind, you'll know!). Those applications probably started out, like Mathcad, in the DOS era. However, this description certainly does not apply to Mathcad because it has—again in my opinion—more successfully taken to the graphic environment than any other mathematics application. It has a completely graphic interface and is truly WYSIWYG (what-you-see-is-what-you-get). This is one of the reasons it is my most-used and best-liked computer application (with the runner-up spot going to the spreadsheet).

There are lots of other reasons why I prefer Mathcad. For example, it is especially good at handling units. My experience with engineering students (including graduate students) over the years has taught me that even the best of them have a somewhat casual attitude to units. (Question: "What's the acceleration of gravity?"

Answer: "32.2"). Mathcad will automatically work with the units for you and give you the answer in any units you wish. It might appear that this is a bad thing, similar to giving a calculator to someone who is innumerate, but actually it's not. In my experience with teaching Mathcad, students get a better understanding for units after having Mathcad as a tool. This is because if a student tries to do something using units that is improper, Mathcad will "flag" that error, forcing the student to check his/her own use of units. For example, if a student inadvertently defines a mass in, say *lbf* rather than *lb*, sooner or later Mathcad will compute something that will have units that the student will recognize to be incorrect. In this way Mathcad will reinforce correct usage of units, while at the same time eliminating all of the drudgery of unit conversions (from, for example, a density expressed in lb/ft^3 to one in kg/m^3).

To their credit, the people at PTC have continued to develop Mathcad at a healthy pace. They seem to release a new version about once every twelve to eighteen months, and each new version genuinely has major improvements. For example, Mathcad 14 has a much-improved symbolic math engine that is more stable and powerful than before. Mathcad's graphing capabilities seem to get better and better, too. While 2D graphing is improving, it is perhaps still a little more powerful in the spreadsheet; 3D graphing in Mathcad is terrific. The latest versions of Mathcad are also incorporating more powerful programming capabilities.

This is the second edition of this book. I have made wholesale changes to the arrangement of the material. For example, I decided to place descriptions of how to graph in Mathcad in an appendix, because no matter what problem you're analyzing in Mathcad you'll probably end up needing to graph something. Having a free-standing appendix on graphing, therefore, seems to make sense; you can refer to it at any point to learn how to graph what you are interested in, without necessarily having to have read previous material.

I'd like to thank a number of people for their help as I prepared this book. Katie O'Keeffe-Swank at Lachina Publishing Services was responsible for making sure I remained on track and for "lubricating wheels" as necessary. Bill Stenquist at McGraw-Hill was also a terrific wheel lubricator, especially for coordinating our efforts with the people at PTC. Katie and Bill are both a pleasure to work with, being simultaneously professional and personable, and were my main support. Of course, I remain full of admiration for the Mathcad people at PTC for producing a marvelous piece of software. At PTC, I'd like to thank Dave Wakstein for getting my Mathcad 14 up and running. I'd also like to thank in advance the reader for updating me (at philip.pritchard@manhattan.edu) on any errors that you may find, and for suggestions for further improvements.

Dedication

To my wife, Penelope, for her support and encouragement in my effort to complete this book on schedule.

About the Author

Philip J. Pritchard received his Ph.D. in engineering mechanics from Columbia University in 1987. He has been a faculty member of mechanical engineering at Manhattan College since 1981, where he teaches undergraduate and graduate courses in thermodynamics, fluid mechanics, and analysis and numerical methods, and has four times been the recipient of the teacher of the year award from the Manhattan College student section of Pi Tau Sigma. He writes textbooks and publishes and presents papers in the area of engineering education, specifically in the use of the personal computer for doing engineering analysis.

What Is Mathcad?

Mathcad: The Good and the Bad

Mathcad is an amazingly useful, powerful, and flexible mathematics computer application for use by engineers, scientists, and students. It can perform many (often all) of the calculations that these types of number crunchers need to do, from the most basic (e.g., converting power in kW to hp), to the most sophisticated (e.g., solving partial differential equations). There are other applications that can also do powerful computations, including Excel, MATLAB, Maple, and Mathematica, as well as programming applications such as Visual Basic (VB) with which you can write programs to do your computations. Some people prefer one or more of these to Mathcad; having used all of them, I think Mathcad has by far the best combination of features for most people. When I do an analysis, I occasionally use Excel and VB, but usually turn first to Mathcad (and I'm not the only one: recently engineering giant Rolls Royce started moving to Mathcad from Excel).

Some of the features that make Mathcad unique are:

1. It is a 100% graphical user interface. Everything you type in Mathcad, from text, to equations, to graphs, is an object that can be dragged and dropped anywhere on the worksheet (the working region of Mathcad). This allows you to rearrange your work very easily. Fig. 1.1 shows a simple workbook; Fig 1.2 shows the same one after we have tidied up—we dragged and dropped several pieces of text and equations, and the graph.

2. Mathcad does not require learning special code. You type expressions pretty much as you do when writing by hand; in Fig. 1.2 we typed the left side of the four equations just as you might expect, with virtually no memorization of special keystrokes. We also typed the right side of the first equation (defining the function $f(\theta)$). Mathcad generated the results of the other three equations.

3. It is live. In Fig. 1.2, if we edit the function $f(\theta)$ (e.g., changing it to $e^{-\theta} cos(\theta)$), the three results, and the graph, instantly change (e.g., the numerical integral immediately becomes 0.604).

4. Mathcad takes care of units. It knows all the major units (starting with SI and working on down), as well as many non-standard units (e.g., Stokes for

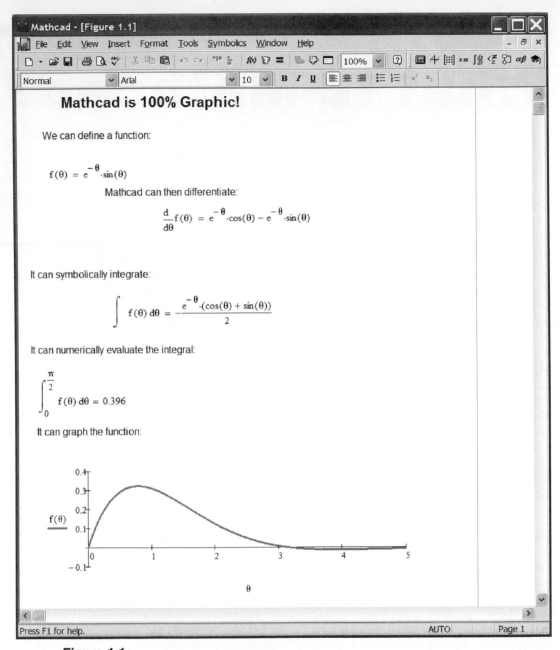

Mathcad is 100% Graphic!

We can define a function:

$$f(\theta) = e^{-\theta} \cdot \sin(\theta)$$

Mathcad can then differentiate:

$$\frac{d}{d\theta} f(\theta) = e^{-\theta} \cdot \cos(\theta) - e^{-\theta} \cdot \sin(\theta)$$

It can symbolically integrate:

$$\int f(\theta)\, d\theta = -\frac{e^{-\theta} \cdot (\cos(\theta) + \sin(\theta))}{2}$$

It can numerically evaluate the integral:

$$\int_{0}^{\frac{\pi}{2}} f(\theta)\, d\theta = 0.396$$

It can graph the function:

***Figure 1.1:
A Mathcad
worksheet before
arranging***

viscosity), and can handle mixed units. Fig. 1.3 shows an example, where we defined the length, width, and height of a box in m, in, and cm, respectively, and asked Mathcad to compute the volume in m^3, ft^3, in^3, and gal. Mathcad did all the work for us!

5. Mathcad is very well integrated with the internet and with other computer applications. You can save a worksheet in html format, and join free Mathcad Web User Forums; you can make Mathcad talk to (e.g., exchange data with, or control) applications such as Excel, MATLAB, and LabVIEW; and you can use controls such as radio buttons and checkboxes, like those in Visual Basic.

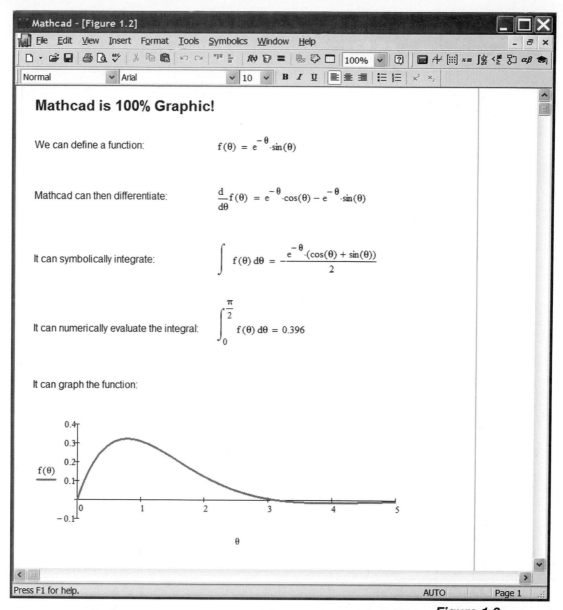

Other applications have some of these features—for example, Excel has elements of all of the above items (yes, even units) —but none of them are as powerful as Mathcad.

That's the good news; there is some less good news. Mathcad works in some ways that make the initial learning curve a little steep:

1. It's pretty easy to type an equation into Mathcad, but when you *edit* an equation Mathcad will do things to the equation that will (until you have some experience with it) seem a bit confusing. Beginners sometimes just delete the equation and start over.

Figure 1.2:
The same Mathcad worksheet after arranging

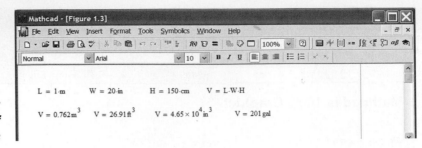

2. Mathcad can do so much without requiring knowledge of special code, but a typical keyboard has fewer than 50 keys. Hence, Mathcad has some special uses for some keys you need to remember. For example, when you type the equals sign (=) Mathcad sometimes produces on the screen an equals sign, but sometimes it produces colon-equals (:=) instead (we'll see the rhyme and reason for this later)!

3. Mathcad is brilliant at not letting you do something that's incorrect. For example, if you ask Mathcad to add 5 kg and 10 m, it will produce an error warning. This doesn't sound like a bad thing, but sometimes people want to shoot the messenger!

One of the goals of this book is to help you master these tricky parts of Mathcad; within a few hours, you'll become a Mathcad power user!

A final comment before we start a detailed description of Mathcad: do not be under the impression that it is a lightweight analysis package compared to its competitors. A lot of engineers and scientists who tried Mathcad years ago dropped it because it used to have some severe limitations. For example, it could only handle very small matrices; it didn't have powerful numerical methods built-in; and it couldn't be programmed. Today Mathcad is tremendously powerful (none of the three examples is true now); it is widely used, for example, by Boeing, Lockheed Martin, and Universal Studios; and *Desktop Engineering* magazine recently announced it was Readers' Choice Product of the Year.

1.2

The Mathcad Interface

It's a good idea to have Mathcad running while using this book. Figure 1.4 shows a typical setup (among other things, we closed the Trace Window): it has a Menu Bar, Standard Toolbar, and Formatting Toolbar.

Mathcad has a unique, and all-important, Math Toolbar. If your Mathcad doesn't show it, you can find it by clicking on *View . . . Toolbars . . . Math* (and of course you can move it, or any other toolbar, around, or place it on any boundary of the window). Clicking on each of its icons opens the toolbars shown in Fig. 1.5. We'll find these very useful for getting quick access to powerful math. Some of these collections are obvious (e.g., the Calculator and Calculus

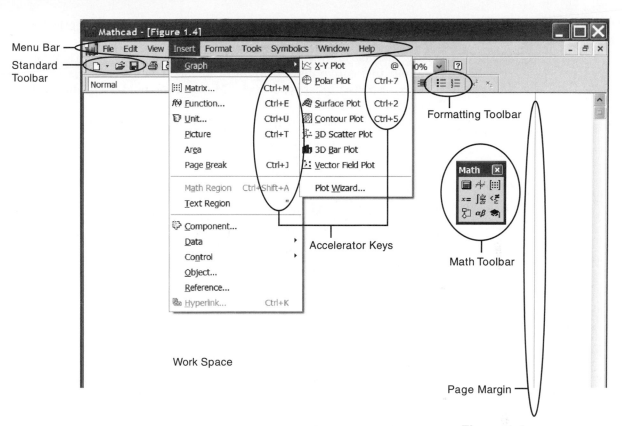

Menu Bar

Standard Toolbar

Formatting Toolbar

Accelerator Keys

Math Toolbar

Page Margin

Work Space

**Figure 1.4:
The Mathcad
interface**

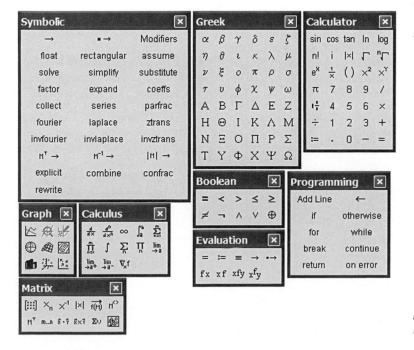

**Figure 1.5:
Useful toolbars**

collections), some less so (e.g., the Symbolic and Boolean collections); we'll eventually use most of them.

Fig. 1.4 also shows some typical accelerator key information; for example, to create a matrix in Mathcad you could click on *Insert . . . Matrix*, or instead just type **Ctrl + M** (meaning hold down the **Ctrl** key while typing **M**).

We can summarize the ways to do something in Mathcad:

1. Use the menu item (e.g., as shown in Figure 1.4 for an X-Y Plot click *Insert . . . Graph . . . X-Y Plot*).

2. Click an icon (e.g., for the graph, click the ⬓ icon in the Graph Toolbar).

3. Use an accelerator key (e.g., for the graph type @ in any empty region).

With these three methods available, you should never be confused about how to create something in Mathcad. Which you use is entirely a matter of personal preference.

The last element of Fig. 1.4 worth mentioning is the Work Space. This is where we'll do all of our work; it's like a white board or sheet of blank paper, but better, because everything (an equation, text, a graph, etc.) we "write" here is an object that can be dragged around! The vertical line defines the page margin: only things that are to the left of it get printed; anything to the right of it will get printed as a later page. (As we already mentioned, we closed the Trace Window—we won't be using it in this book.)

Mathcad Basics

2

The goal of this chapter is to get you up and running; after completing it you should be able to do quite a lot of basic math with Mathcad. It's best to learn by solving a specific problem, so let's consider the following:

Example 2.1: Projectile Analysis A projectile of mass 10 kg is fired at a speed of 50 m/s, at an angle with the horizontal of 30°. Find the location after 1 s (ignore friction). What was the initial kinetic energy?

The completed worksheet is shown in Fig. 2.1.

We will generate this worksheet and in doing so demonstrate:

✓ Entering and editing equations
✓ Using built-in, and creating user-defined, functions
✓ Entering text
✓ Formatting a worksheet

After this chapter we will *not* use a roundabout route to solve problems, but in this chapter we want to carefully illustrate each of the above points. Our first step is to answer the problem with as little detail as possible. Our crude first effort will look like Fig. 2.2.

Let's get started. Open Mathcad, and you'll have an empty worksheet. We want to generate the equations shown in Fig. 2.2. We will step through this, but if you make a typo (or part of the equation goes red), it's probably easier to delete and start over for now; we will learn how to edit correctly in Section 2.2.

2.1
Entering Equations

When we use Mathcad we need to always think about input and output. If we input something, we are defining something for Mathcad; if we get an output, we are asking Mathcad to give us a result. This leads to:

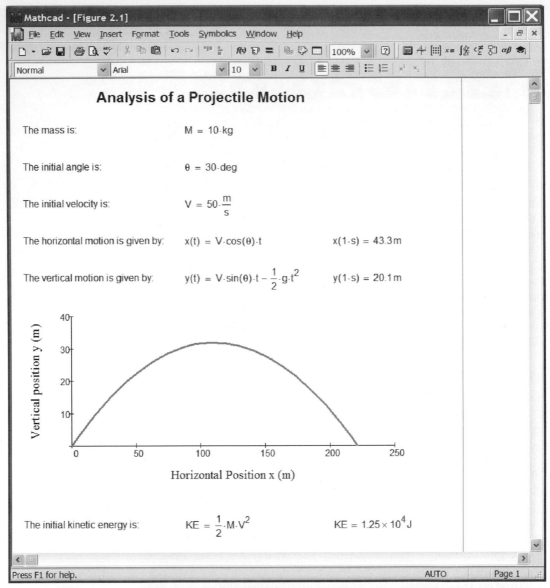

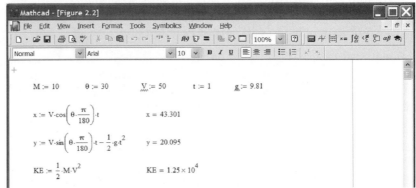

Figure 2.1:
The finished
worksheet for
Example 2.1

Figure 2.2:
The initial worksheet
for Example 2.1

Mathcad's Four (Well, Maybe Five) Equals Signs

1. The *assignment* equals sign. This is used to *input* or *define* something for Mathcad. To create it you type **:** (colon) or use the icon in the Evaluation Toolbar (but see the discussion on variables on pages 9 and 10).

  You can go ahead and type the first line of Fig. 2.2. Click on the worksheet and type in the equation for M. To finish an equation, just press **Enter** (wherever you are in the equation!) or use the arrow keys ($\leftarrow$, $\uparrow$, $\rightarrow$, and $\downarrow$) on the keyboard, or click on an empty space with the mouse. Repeat for the other four terms. For θ you can use the Greek Symbol Toolbar. For any Greek letter you can either use the toolbar or instead type the equivalent Roman letter and immediately press **Ctrl + G** (meaning hold down the **Ctrl** key while typing the **G** key); in this example you'd type **q** then press **Ctrl + G**).

2. The *evaluation* equals sign. You simply type **=** or use the icon in the Evaluation Toolbar (but see the discussion on variables below). This is used to get *output* or ask Mathcad to provide a result.

  We will use this after we have defined x, y, and KE in Fig. 2.2.

3. The *priority* equals sign. To get it type **~** (the tilde key) or use the icon in the Evaluation Toolbar. This is sometimes used to *input* or *define* something instead of using the assignment equals sign. Mathcad always evaluates worksheets starting from the top left and working down toward the end of the worksheet. If you want to want to define something at the end of the worksheet even though you used it earlier on, you define it using the priority equals sign. Mathcad *always* evaluates equations with priority equals first, *then* it goes on to evaluate all other equations.

4. The *Boolean* equals sign. To get it type **Ctrl + =** or use the icon in the Boolean Toolbar. This *input* symbol is used in equations when we wish to symbolically or numerically solve one or more equations for an unknown or unknowns.

  We will use this a lot in Chapter 3.

5. *Symbolic* equals sign. To get it type **Ctrl + .** or use the icon in the Evaluation Toolbar. This *output* symbol is used when we want Mathcad to provide a symbolic (i.e., a result in terms of symbols) instead of a numerical result.

  We will use this a lot in Chapter 5. We called this an equals sign even though it really doesn't look like one!

If you succeeded in typing the first line of Fig. 2.2, you'll notice that variables V and g both have a squiggly line under them. We need to understand how Mathcad handles variables to explain this.

Variables: Built-in and User-Defined

We create or define variables in Mathcad using the assignment equals, as shown in Fig. 2.2. This worked fine for M, θ, and t but led to the squiggly lines for V and g. The squiggly lines are a warning: Mathcad is telling us that we are *redefining* things that Mathcad already knows about and uses as built-in variables. In this case Mathcad already uses V for volts and g for gravity. You are free to redefine any built-in variable for your own use, but you should be wary of doing so because you may get unexpected results. (You can switch off display of these lines if you wish; see menu item *Tools . . . Preferences . . . Warnings*.) A classic mistake is to use m for the mass of something; it's better to use M for the mass of something. Using m for mass just defines away the use of m for meters! The single letter built-in variables are shown in Fig. 2.3. There are many other built-in variables; three examples are shown in Fig. 2.3. This is not as bad as it sounds; for example, in a dynamics problem, you might use V for velocity, replacing the default unit V for volts; in a dynamics problem, you're probably not using volts anyway. Of course, you may not wish to use any units at all in your calculations (as in Fig. 2.2); in which case feel free to redefine m, g, and so on.

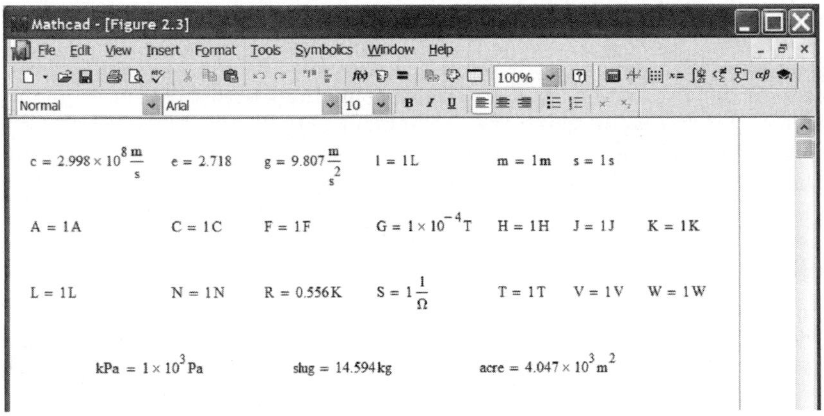

Figure 2.3: Some built-in variables

This raises an interesting point about the assignment (input, :=) and evaluation (output, =) equals signs: Mathcad tries to help you do the right thing! If you inadvertently type = instead of : when assigning a variable, Mathcad automatically generates the assigned equals for you if the variable does not already exist. We can summarize this:

Type this	**x:**	to get	$x:=$	
Typing this	**x=**	produces	$x:=$	if x is not yet defined
Typing this	**x=**	produces	$x=$[value of x]	if x is already defined (user-defined or built-in)

If you get a squiggly line (as in Fig. 2.2), you can click on the expression, and Mathcad will tell you what the warning is about. It's then up to you as to whether you wish to change the equation or leave it.

We are ready to do the rest of Fig. 2.2. To do this we need to discuss built-in functions.

Built-in Functions

Mathcad has an amazing number of built-in functions. To see them all, use menu item *Insert . . . Function* or click on the 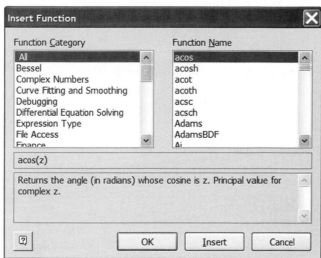 icon on the Standard Toolbar or type **Ctrl + E**. You'll get the window shown in Fig. 2.4.

Figure 2.4: Mathcad's built-in functions

To insert a function into an equation you're working on, you can just find it in this window and click *Insert*. A quicker alternative, which you'll almost always use (at least for common functions), is to simply type the function name into your equation.

Let's type in the rest of Fig. 2.2, starting with the equation for x. First, cancel the Insert Function window if it's still open. To create the rest of the equation just type as if Mathcad were a work processor such as Word, except for the following important tips:

Tips for Entering Equations

1. To define a variable, type **:** (or you can type **=** if the variable has not yet been defined by you or Mathcad).

2. To multiply, type *****; to divide, type **/**; for an exponential type **^**.

3. To create Greek symbols, use the Greek Symbol Toolbar or type the Roman equivalent and immediately type **Ctrl + G** (e.g., for π type **p** and then **Ctrl + G**).

4. Do *not* use the **Spacebar** anywhere (*except* see important item 7 below); variable names and functions cannot have spaces. If you do use the **Spacebar**, you might convert your equation into a text region!

$$x := V \cdot \cos(\theta \pi]$$

5. If you use parentheses, you can type the closing parenthesis at any point and Mathcad will automatically enclose everything between the opening and closing pair (even if you happen to be in the denominator of an expression).

6. To exit an equation press **Enter** or use the mouse or arrow keys (←, ↑, →, and ↓) on the keyboard.

7. Pay attention to the editing bars (the *underline*, the horizontal editing line, and the *insertion* line, the vertical editing line, as shown in Fig. 2.5. They tell you what part of the equation you're working with. In the figure, we are about to divide π by 180.

This is where the **Spacebar** comes in handy: you can use it to expand or contract the editing bars. For example, in Fig. 2.5, pressing the **Spacebar** would expand the editing lines to include θ and π (and if you now divided, both would be in the numerator); pressing again and again would expand the lines to include more and more of the equation, until you cycle back to Fig. 2.5. Important uses for the **Spacebar** are to get out from under the denominator, or out of an exponent (power), of an expression. (This feature is so important we will repeat it when we discuss editing an equation in Section 2.2!)

8. Make sure a variable is defined (built-in or user defined) before asking Mathcad to compute using it, otherwise you will get an error! This seems obvious, but sometimes is a bit subtle. Mathcad reads and computes worksheets from top left to bottom right (with the exception of the priority equals discussed above). Figure 2.6 shows what happens if an evaluation is moved even slightly above its defining equation.

$$x := V \cdot \cos\left(\theta \cdot \frac{\pi}{180}\right) \cdot t \quad x = 43.301 \qquad x := V \cdot \cos\left(\theta \cdot \frac{\pi}{180}\right) \cdot t \qquad \boxed{x = \blacksquare\blacksquare}$$

This variable is undefined

(a) Equations in line   (b) Second equation slightly above first

Figure 2.6:
How Mathcad
reads equations

9. To move a group of equations (or even just one equation), wipe over them with the mouse so you get dashed lines around each equation, and then use the arrow keys (←, ↑, →, and ↓) on the keyboard to move the equations in the direction you want. Alternatively: To move a single equation, click on it, and then move the mouse to the box around the equation. You should get a small black hand, ✋, at which point you can click and drag the equation. To move more than one equation at a time, wipe the mouse over them (or click on one equation and then hold down the **Ctrl** key while clicking on the others); then move to the edge of any of the dashed boxes around the equations, and you'll get the hand again, ready for you to move the equations around. Finally, to line up selected equations you can use the ⬚⬚ icons on the Standard Toolbar.

You should now be able to get the worksheet to look like Fig. 2.2. Note that Mathcad computes *x*, *y*, and *KE* for you! We'd now like to edit this worksheet to improve it and end up with the final worksheet shown in Fig. 2.1. To do this we need to do some editing. Editing an equation requires some practice; we discuss this in detail next.

Exercises

Notes: For some of these problems, you will find Mathcad's built-in Reference Tables very helpful. Click menu item *Help . . . Reference Tables.* For these exercises it's probably a good idea, until you become a "power user," to solve each one in a separate worksheet—otherwise, a variable name used in several different exercises in the same worksheet may lead to some confusion.

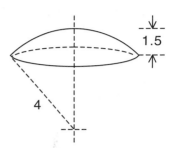

Figure E2.5, E2.6, E2.7

2.1 Find the surface area (including end surfaces) of a cylinder of height 5 and diameter 5.

2.2 Find the volume of the cylinder of Exercise 2.1.

2.3 Find the surface area (including the end surface) of a right circular cone of height 5 and diameter 5.

2.4 Find the volume of the cone of Exercise 2.3.

2.5 Find the surface area of upper surface of the cap shown.

2.6 Find the surface area of the cap shown if it is a solid, (i.e., include the area of the disk underneath the cap).

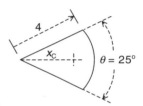

Figure E2.8, E2.9, E2.10

2.7 Find the volume of the cap shown.

2.8 Find the area of the "pizza slice" shown.

2.9 Find the total length of the perimeter of the "pizza slice" shown.

2.10 Find the centroid x_C of the "pizza slice" shown (the centroid is the point at which the object could be balanced on a pinhead).

2.11 You invest $1000 at an annual interest rate of 3.5%. Find the value of your investment after 10 years, assuming (a) the interest is added at the end of each year (in other words, it is compounded annually), (b) it is compounded monthly, and (c) it is compounded daily. Use the following formula:

$$F = P \cdot \left(1 + \frac{i}{k}\right)^n$$

where P is the principal (initial investment), i is the annual interest rate (expressed as a decimal), k is the number of compounding periods a year (assume 12 months a year and 365.25 days a year), and n is the total number of periods. Hint: Do not use units. Solve (a), then copy and paste and edit as necessary for (b) and (c).

2.12 The temperature of a cooling body is given $T = T_0 e^{-\frac{t}{\tau}}$ where $T_0 = 500$ and $\tau = 60$. Find T when $t = 25$.

2.13 For Exercise 2.12, manually solve the formula for t (we'll see in Chapter 5 that Mathcad could do this for us!). Then find t at which $T = 100$ (use the same T_0 and τ).

2.14 Evaluate the following: (a) cosh(0), (b) cos(π/3), (c) $J_0(2)$ (zeroth-order Bessel function of the first kind), (d) log(e), (e) 1/ln(10).

2.15 Evaluate the following: (a) $\cosh^{-1}(0)$, (b) $\cos^{-1}\left(\dfrac{\sqrt{3}}{2}\right)(°)$, (c) $P_2(0)$ (Legendre polynomial of degree 2), (d) $\mathrm{erf}(1)$, (e) $\mathrm{erfc}(1)$.

2.2

Editing Equations

New users usually have the greatest difficulty with editing equations; *it is the single most difficult skill to master in Mathcad.* As we mentioned in Section 2.1, when you first create an equation, you type pretty much as you would in a word processor; the equation gets built much as you expect, with the help of the nine hints in Section 2.1. *Editing* an existing equation takes some practice; this is because, after all, when you type equations in Mathcad, you're really giving Mathcad computing instructions; equations are not just pretty, they actually compute live math!

We will practice editing by working with our worksheet as in Fig. 2.2. First let's list some tips for editing.

Tips for Editing Equations

1. To enter an equation for editing, either use the mouse to click on it, or use the arrow keys (←, ↑, →, and ↓) on the keyboard to navigate into it.

2. As in the tips for creating an equation, pay close attention to the editing bars (the underline and insertion lines); they tell you what part of the equation you're about to modify. This is where the **Spacebar** comes in handy: you can use it to expand or contract the editing bars. Pressing **Spacebar** again and again expands the editing lines to include more and more of the equation, until you cycle back to your original selection. Important uses for the **Spacebar** are to get out from under the denominator or out of the exponent (power) of an expression. You can also use the left and right arrow keys (← and →) for moving around in the equation; the up and down arrows (↑ and ↓) leave the equation, unless you're in a denominator or numerator, respectively. The **Insert** key can be used to repeatedly change the insertion line from the left to the right of the underline line.

3. Using the **Delete** and **Backspace** keys:

(a) If the editing lines enclose only one variable or number (or the insertion line is somewhere in the middle of the underline line), **Backspace** deletes whatever is to the *left* of the insertion line, and **Delete** deletes whatever is to the *right* of the insertion line. *What gets deleted could be a single letter, a digit, or an operator* (such as multiplication or division). See Fig. 2.7, first row.

(b) If the editing lines are larger, and enclose more than one variable or number, it's more complicated. If the insertion line is on the *right*: **Backspace** highlights in black *everything* enclosed between the editing lines (to delete, press **Backspace** again—see Fig. 2.7, second row); **Delete** works as in item (a) above. If the insertion line is on the *left*: **Backspace** works just as in item (a) above; **Delete** highlights in black *everything* enclosed between the editing lines, to let you know what is about to be deleted (to delete, press **Delete** again—see Fig. 2.7, third row).

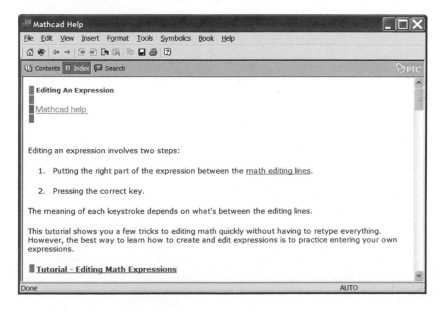

In this situation:	**Backspace** does this:	and **Delete** does this:
$x := V \cdot \cos\left(\theta \frac{\pi}{180}\right) \cdot t$	$x := V \cdot \cos\left(\frac{\pi}{180}\right) \cdot t$	$x := V \cdot \cos\left(\theta \square \frac{\pi}{180}\right) \cdot t$

In this situation:	**Backspace** does this:	then this:
$x := V \cdot \cos\left(\theta \cdot \frac{\pi}{180}\right) \cdot t$	$x := V \cdot \cos\left(\theta \cdot \frac{\pi}{180}\right) \cdot t$	$x := V \cdot \cos(\theta \cdot) \cdot t$

In this situation:	**Delete** does this:	then this:
$x := V \cdot \cos\left(\theta \frac{\pi}{180}\right) \cdot t$	$x := V \cdot \cos\left(\theta \cdot \frac{\pi}{180}\right) \cdot t$	$x := V \cdot \cos(\theta \cdot) \cdot t$

Figure 2.7:
Examples of using
Backspace and Delete

4. To completely delete an equation you can: click on it and press **Spacebar** to completely enclose it, then **Delete** or **Backspace** as appropriate; or wipe over a region enclosing the equation or equations (each equation will then have a dashed line around it) and press **Delete**.

If you can master these editing tips you'll be well on the way to becoming a Mathcad expert. It takes a bit of practice, but you'll eventually develop a sense of the logic behind Mathcad equation editing.

Figure 2.7 also illustrates another Mathcad feature: *placeholders*. These are the little black squares that appear when you delete something and an equation is incomplete. Note also that when you delete an operator you get a small hollow square.

There is more information on editing in Mathcad's Help menu; an example page is shown in Fig. 2.8.

Figure 2.8:
Help on editing an
equation

Let's practice editing by improving the worksheet of Fig. 2.2 by introducing two new topics: user-defined functions and units.

Exercises

Note: For these exercises, type into a blank Mathcad worksheet the "Before" expression, then practice editing to get the "After" expression. Then edit the "After" to get back to the "Before."

2.16 Before: $x^2 + 1$ After: $e^{\sqrt{x^2+1}}$

2.17 Before: x^2 After: $(x + 1)^2$

2.18 Before: $\dfrac{(x + y)}{(x - y)}$ After: $(x + y)(x - y)$

2.19 Before: $\sin(x + y) \cdot e^x$ After: $\dfrac{\cos(x + y)}{e^x}$

2.3

User-Defined Functions

In Fig. 2.2 (repeated below for convenience) x and y are simply *variables*; we want to make them into user-defined *functions* (as in Fig. 2.1). A variable is something that is assigned a value; a function is something that has one or more arguments that are used in assigning a value to the function. For example, $\sin(\theta)$ and $\cos(\theta)$ are built-in functions that need one argument, θ; $\text{atan2}(x, y)$ is a built-in function that needs two arguments, x and y (it computes the angle of a right triangle). *One big difference between a variable and a function is that you can define a function even though its arguments are not yet defined.* For example, we want x and y to be variables with time t as argument. To illustrate the difference between a variable and a function, let's delete the equation for t. We end up with the worksheet of Fig. 2.9.

Mathcad shows the equations for x and y with red t's, and clicking on one of those t's tells you the error: we have tried to compute x and y with something that is not yet defined (so they fail to compute)!

We can fix this by making x and y *functions* of t (click or use the arrow keys to get the editing lines after the x of the x equation; type **(t)**; press **Enter**; repeat for the y equation). After editing, we get Fig. 2.10 (note that depending on your version of Mathcad you may get something different than $x = f(\text{any1}) \rightarrow \text{any1}$ and $y = f(\text{Unitless}) \rightarrow \text{Unitless}$).

You can delete these x and y variable evaluations because we will now be evaluating functions $x(t)$ and $y(t)$ instead. For example, we can evaluate x and y for $t = 1$ by typing **x(1)=** and **y(1)=** (try it!). This demonstrates that when *defining* functions you don't need to have previously defined its argument or arguments (in this case

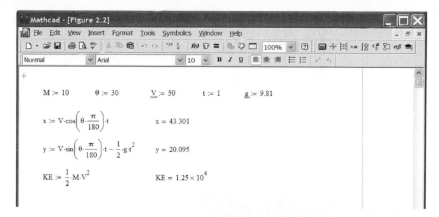

Figure 2.2 (Repeated):
The initial worksheet
for Example 2.1

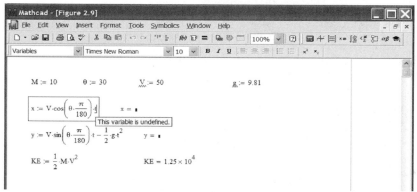

Figure 2.9:
Variables with
undefined arguments

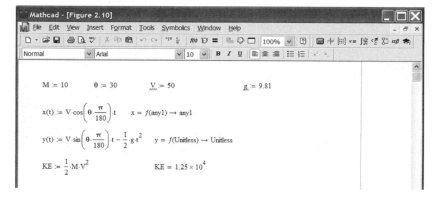

Figure 2.10:
Defining functions

t), but to *evaluate* functions you must use known values for the arguments (just as, say, tan(θ), a built-in function, needs a value for θ before it can be evaluated).

We are almost ready to get to work with the next task for Example 2.1—working with units—but first let's practice working with functions of two variables. Figure 2.11 shows two user-defined functions *f* and *g* (Note that Mathcad puts a squiggly line under *g* to tell us we've just defined away *g*, acceleration of gravity!). Each of these is a function of two variables, *x* and *y*, and they are

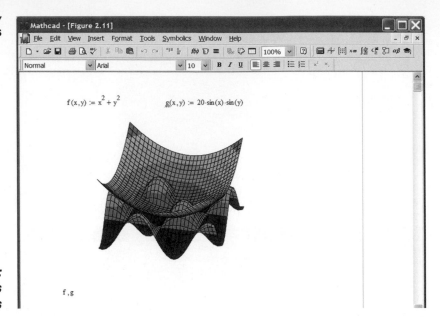

Mathcad - [Figure 2.11]

File Edit View Insert Format Tools Symbolics Window Help

Normal   Arial   10   **B** *I* U

$$f(x,y) := x^2 + y^2 \qquad g(x,y) := 20 \cdot \sin(x) \cdot \sin(y)$$

f,g

Figure 2.11:
*A plot of functions
with two arguments*

plotted. If you wish to practice this, start a blank worksheet, type the equations for *f* and *g*. Then, click somewhere well below the equations; on the menu bar click *Insert . . . Graph . . . Surface Plot*; in the black placeholder on the graph type **f,g** and press **Enter**. You'll get a crude wire-frame surface plot. If you want to make the graph pretty, double click on it for all kinds of ways of tweaking the appearance. Have fun!

Let's get back to our original worksheet (Fig. 2.10). We are ready to learn about units.

Exercises

Note: For these exercises, after defining the function, evaluate it at $t = 0$, 1, and 5 or $(x, y) = (0, 0), (1, 1), (1, 5), (5, 1),$ and $(5, 5)$, as appropriate. In addition, use a default QuickPlot for each one. (For an X-Y Plot type **@** or click on the ⬚ icon in the Graph Toolbar; type **t** in the middle horizontal placeholder and **f(t)** in the middle left placeholder. For a Surface Plot type **Ctrl + 2**, or click on the ⬚ icon in the Graph Toolbar, and type **f** in the placeholder. (If you want to get fancy, you can double-click on a graph and play around with formatting it!)

2.20 $f(t) = t^2 - 6t + 5$

2.21 $f(t) = e^{-\left(\frac{t}{5}\right)^2} \cos(t)$

2.22 $f(t) = \cos\left(\left(\frac{\pi}{8}t\right)^2\right)$

2.23 $f(t) = \dfrac{\sin(t)}{t}$

2.24 $f(x, y) = e^{-\frac{x}{5}}\cos(y)$

2.25 $f(x, y) = J_0\!\left(\dfrac{x^2 + y^2}{5}\right)$ (J_0 is the zeroth-order Bessel function of the first kind.)

2.26 $f(x, y) = e^{-\frac{x^2 + y^2}{20}}$

2.4

Working with Units

Mathcad's handling of units is one of its very best features because it's so powerful and so easy! Other computer applications can't handle units, or if they do, they handle them pretty crudely (e.g., Excel).

We'll get practice in units by making our worksheet (Fig. 2.10) look like Fig. 2.1. First some tips:

Tips for Using Units

1. To insert a unit in an equation do any of the following:

    (a) Just type it like you would any other variable!

    (b) Click on the 🔄 icon in the Standard Toolbar. You'll get the window shown in Fig. 2.12. Typing any letter will make the Dimension list shoot to that letter (e.g., typing **L** shoots to Length), and you can then select which unit of length you want (e.g., for dimension length you get a Unit list starting with Angstroms).

    (c) Click on *Insert . . . Unit* to get the window shown in Fig. 2.12.

    (d) Type **Ctrl + U** to get the window shown in Fig. 2.12.

2. When you *evaluate* something Mathcad will give the answer in default units (SI). If you wish to change the units displayed: click on the equation to get the black units placeholder, and then use any method above for inserting there a unit or units (or double-click on the black placeholder to instantly get the Insert Unit window with the correct dimension already preselected), and then exit the equation. Mathcad will automatically recompute the result and display the answer with the units you chose. Any "leftover" units needed to complete the computation will also be displayed.

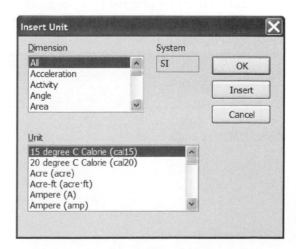

Figure 2.12:
The Insert Unit
window

3. Mathcad works with SI as default units. You can change this to, for example, U.S. units, by clicking menu item *Tools . . . Worksheet Options . . . Unit System*.

4. Regardless of default units, Mathcad will compute with default and non-default units. For example, you can mix SI and U.S. units in a calculation and Mathcad will take care of all the unit conversions and give the answer in default units.

5. You can insert units into an equation you're creating using the methods of item 1 above without using the multiply sign first, but when you exit the equation you won't see the multiply sign displayed (although the equation still works); however, *it's preferable to use the multiply sign because then you'll see it when you exit the equation.*

6. You can define your own units (just like you would define a variable).

7. Be wary of defining away important units (see Fig. 2.3).

8. You are not *required* to use units; a Mathcad worksheet can just be pure math.

9. Mathcad will show errors in units. For example, Fig. 2.13 show what happens if you do something improper. It will allow some "errors" (see the discussion of Fig. 2.14).

$$M := 10 \cdot kg \qquad W := 20 \cdot N$$

Figure 2.13:
An example of
improper unit use

$$a := M + W$$

This value has units: Force,
but must have units: Mass.

Let's use these tips in using units in our example worksheet. Insert kg and deg into the equations for M and θ. Note that, like other applications and scientific calculators, Mathcad assumes angles are in radians (that's why we had the $\pi/180$ factors in the equations for x and y); here we provided θ in degrees. You can now delete the $\pi/180$ factors and get a worksheet like Fig. 2.14.

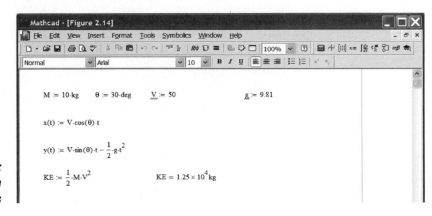

Figure 2.14:
The worksheet with
some units

In Fig. 2.14 we see that the *KE* was computed, but with wrong units! This is because we haven't fully used units in the worksheet (*M* is in kg, but *V* is just a number, so $\frac{1}{2}MV^2$ is in kg). The point is, be careful with units: if you make improper use of units Mathcad will flag you (as in Fig. 2.13), but sometimes you can do something wrong with units that still computes (as in Fig. 2.14).

Let's finish the units for this example: insert units for velocity *V* (m/s); delete *g* and let Mathcad use its built-in value (*evaluate* *g* instead by typing **g=**); and evaluate *x* and *y* after 1 s. Note that you *must* be consistent with units: when you evaluate *x* and *y*, the argument (*t*) must have units of time, not just a number (actually using just a number works for *x* and not *y* here because the *x*

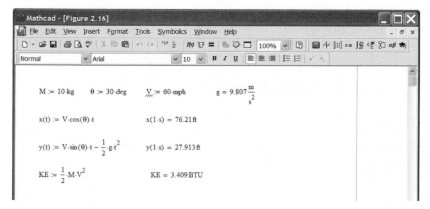

Figure 2.15:
The worksheet with
completed units

function uses *t* in an especially simple way). Try using just 1 in the evaluation of *y*, and you'll get an error.

Now the fun with units begins: we can change any of the input values and units to immediately get new results, and we can also change the units of the evaluations!

Try changing *V* to 60 mph, and then change the evaluation of *x* and *y* to be ft, and *KE* to be BTU. You should get a worksheet like Fig. 2.16. It demonstrates instant recalculation, Mathcad's handling of mixed units, and Mathcad's ability to evaluate in any units you desire.

Figure 2.16:
The worksheet with
mixed units

Exercises

Note: For some of these problems, you will find Mathcad's built-in Reference Tables very helpful. Click menu item *Help ... Reference Tables*.

Note: Be careful with variables and units in these problems! If you do several exercises in the same worksheet, there's a good chance you'll get confusing results (e.g., if you define V as velocity in one problem, you will *no longer* have V for volts as a built-in unit for a later problem). To avoid this, either use a separate worksheet for each exercise or make sure you are aware of this (and type, for example, **V:=volt** for the second problem).

2.27 Suppose you weigh 150 lbf. Find your weight in N and your mass in lb and in kg.

2.28 A cylindrical tank of diameter 25 in and height 3 ft contains kerosene. How many gallons is this? How many liters? How many kilograms? How many pounds?

2.29 A truck's maximum load is 5000 lb. How many copper rods of diameter 1 cm and length 5 m can the truck carry? Note: The Reference Tables give the specific gravity SG of solids; to get the density, use $\rho_{Solid} = SG_{Solid} \cdot \rho_{Water}$. Also, it is common practice to use lb to refer to weight (strictly speaking it's lbf) and also to mass.

2.30 A motor has to be selected for a machine to lift 6000 kg/min a vertical distance of 20 m. Find the minimum theoretical hp required. Note: Work is force × distance, so power is force × distance / time.

2.31 Find the resistance and current in a 100-W bulb plugged into a 120-V outlet. Find the current flowing through and resistance of the bulb. Note: Power consumption is given by RI^2, and by VI, where R is resistance and I is the current; be careful here, because V is a unit in Mathcad, so you probably don't want to use V as a variable name!

2.32 An elephant can weigh as much as 10,000 lbf. Consequently, they will usually walk around a hill rather than over it. Find how many calories are equivalent to an elephant moving uphill just 100 ft. Note: Walking uphill essentially stores potential energy Mgh, where M is mass and h is height. Also, there are two kinds of calories commonly defined, the scientific one, and one used in the diet industry (1000 times the scientific one)—Mathcad uses "cal" and "dcal" for these, respectively.

2.5

Entering Text Regions

Everything you type into Mathcad is a *math region* ... unless you tell it otherwise! How do you tell Mathcad you're entering a *text region*? Let's look at some tips for entering text:

Tips for Entering Text Regions

1. There are several ways to create a text region:

   (a) Type " before typing the expression. When you do this, Mathcad understands that what you're about to type is a text region (you don't need to type ending quotes).

(b) Click on menu item *Insert . . . Text Region*.

(c) Start typing, but make sure you type a space (press the **Spacebar**) before using any other punctuation (such as ., ,, ;, :, (, etc.). When you type, Mathcad assumes you're typing a math region until you type a space, at which point it realizes you're typing a text region.

You'll probably find that method (c) is most convenient, but whichever one you use is a matter of personal preference.

2. To exit a text region don't press **Enter** (this will just start a new line in the text region); you either use the arrow keys or use the mouse to exit a text region.

3. By default a text region only wraps when it reaches the right margin, so it will obscure any math regions that are to its right. There are several ways to fix this:

(a) Before reaching a math region you can press **Enter** to start a new line in the same text region.

(b) Before reaching a math region you can right-click in the text region and select *Properties* (or click on menu item *Format . . . Properties*) and in the window that appears choose *Push Regions Down As You Type*.

(c) At any time you can click and drag on the small placeholder, ❙, on the right boundary of the text region to resize the text box.

(d) You can drag the text box to a new location using the small black hand, 🖑, that appears when you move the mouse over any part of the text box boundary.

(e) You can click on menu item *Format . . . Separate Regions* (but be aware this will separate regions in the entire worksheet).

4. The default style for text regions is Normal (Arial 10 pt.), but you can select other predefined styles such as Title, Heading 1, and Heading 2. You can change the style properties to your tastes (see formatting in Section 2.6).

5. You can insert a live equation (a math region) within a text region. To do so, while creating or editing a text region, click on menu item *Insert . . . Math Region*.

You can now go ahead and type all the text of our example worksheet, as shown in Fig. 2.1; we will tidy up the worksheet by learning a bit about formatting.

2.6

Formatting Your Worksheet

After doing some work on your worksheet, you probably have math and text regions all over the place. To move a region, click on it, and then move the mouse to the box around the region. You should get a small black hand, 🖑, at which point you can click and drag the region. To move more than one region at a time, wipe the mouse over them (or, click on one region; then hold down the **Ctrl** key while clicking on the others); then move to the edge of any of the dashed boxes around the regions, and you'll get the hand again, ready for you to drag. Finally, to line up selected regions, you can use the ⌗⌗ ⌗ icons on the Standard Toolbar.

By default Mathcad creates all math regions using Times New Roman 10 pt. and all text regions using Arial 10 pt. In addition, computed results are by default presented with three decimal places and a threshold of three for changing to exponential format. We have also already seen that Mathcad has a number of different equals signs (Section 2.1).

All of this is fine, but sometimes we want to change these formats. We discuss text and math regions separately:

Formatting Text Regions

Formatting text regions is as easy and almost as powerful as it is in a word processor such as Word. If you're a power user of Word, you know about file templates, styles, and so on. Mathcad has the same kind of power. Obviously, we can't review all the bells and whistles, but the most important feature is the Formatting Toolbar, shown in Fig. 2.17.

Figure 2.17: The Formatting Toolbar

1. To change the appearance of an entire text region, simply choose a different style from the toolbar; in addition, there are icons for left- and right-justifying, and centering.

2. To change the appearance of a selected piece of text within a text region, choose a font and point size, as well as bold, italic, and underline options, from the toolbar.

3. For changing *all* text regions that have the same style in a worksheet, you can modify the style itself. Click on menu item *Format . . . Style* to get the window shown in Fig. 2.18. You can then customize one or more styles (and you can even save these styles as a new document template).

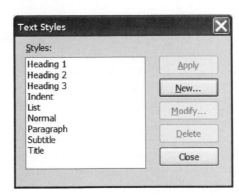

Figure 2.18: The Text Styles window

Formatting Math Regions

Math regions also have styles. To see available styles click on menu item *Format . . . Equation* to get the window shown in Fig. 2.19. There are two obvious styles: those applied to all variables and those applied to all constants (numbers). By default they are both Times New Roman 10 pt., but you can modify them if you wish. You can also define your own math styles: for example, if you work with vectors you may wish to create a Vector style that makes vectors bold.

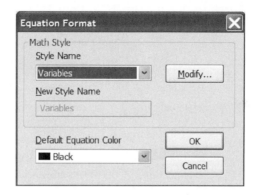

Figure 2.19:
The Equation
Format window

You can also format the results of a computation. Using the Result Format window shown in Fig. 2.20, you have a number of fairly obvious options available to you (to get this window, double-click on a result, or click on menu item *Format . . . Result*). Note that you can set the changes you make as defaults for the entire worksheet.

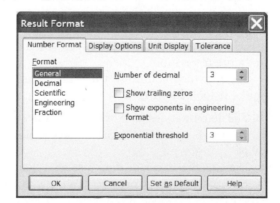

Figure 2.20:
The Result Format
window

We learned that Mathcad has several different equals signs. This is very useful: when you open a worksheet you made a while ago, you can easily see what's been defined and what's been computed, just by looking at the equals signs. On the other hand, if you wish to publish or present your work, you probably want all your equals signs to look standard (=). You can do this by clicking on menu item *Tools . . . Worksheet Options* to get the window shown in Fig. 2.21. Just be aware if you do this change it will no longer be obvious what your worksheet logic is (what you defined, what you evaluated). Note that this window also lets you do things like change the default unit system.

Figure 2.21:
The Worksheet
Options window

Finally, under *Tools . . . Preferences* you can change such things as whether or not Mathcad gives you a squiggly underline if you redefine a variable.

We can now complete the worksheet we've been working on to make it look like Fig. 2.1:

1. The default styles for Normal and Title styles are left unchanged.

2. The math styles for Variables and Constants are both changed to Arial 10 pt.

3. The assignment equals signs display is changed to the regular equals sign.

4. We inserted an X-Y Plot. We will learn about graphing in the Appendix: Graphing, but for now, if you want to do the graph:

(a) On the menu bar click *Insert . . . Graph . . . X-Y Plot*.

(b) Click on each of the three black placeholders on the horizontal axis and type (starting with the one on the left), **0**, **x(t)**, and **250*m**.

(c) Click on the black placeholder at the bottom of the vertical axis and type **0**; on the middle placeholder type **y(t)** (leave the top one blank).

(d) Press **Enter** to exit the graph.

(e) You can move the graph by clicking once on it and moving the mouse to the edge, where you'll get a small black hand, 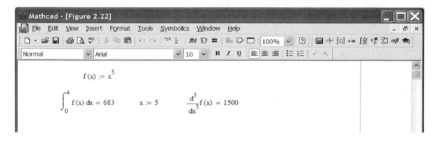 ; use this to move the graph.

(f) You can resize the graph by clicking once on it and then dragging on the small placeholder, ⌐ , on the right, bottom, or corner boundary of the graph.

We are now finished with our example worksheet. You can now save it, print-preview, print it, etc. Before you print, you can check the page setup, change the margins, and the like, much as you would in Word.

The goal of this chapter was to be a self-contained introduction to getting started with Mathcad. The other chapters are designed to give detailed information on specific features of Mathcad (the bells and whistles), but before we proceed we can't resist discussing briefly one of the bells . . .

Mathcad is terrific at performing integration and differentiation. Let's do the following example to illustrate.

A Mathcad "Bell": Calculus

Example 2.2: Integration and Differentiation Evaluate the following: $\int_0^4 f(x)dx$ and (at $x=5$) $\dfrac{d^3f(x)}{dt^3}$, where $f(x) = x^5$.

The solution is presented in Fig. 2.22. To create the integral and derivative, use the Calculus Toolbar. (Do not attempt to type the d's . . . you must use Mathcad's built-in derivative operator!) See how easy that is?

We will have a lot more to say about calculus in Chapter 5.

```
Mathcad - [Figure 2.22]
File  Edit  View  Insert  Format  Tools  Symbolics  Window  Help

Normal          Arial              10    B  I  U

        f(x) := x^5

   ⌠4
   ⎮  f(x) dx = 683      x := 5       d³
   ⌡0                                 ──f(x) = 1500
                                      dx³
```

Figure 2.22:
The worksheet for
Example 2.2

Exercises

2.33 Integrate $f(t)$ of Exercise 2.20 from $t = 1$ to $t = 5$. Also find the slope and second derivative at $t = 0$.

2.34 Integrate $f(t)$ of Exercise 2.21 from $t = 0$ to $t = \pi$. Also find the slope and second derivative at $t = \pi$.

2.35 Integrate $f(t)$ of Exercise 2.22 from $t = 0$ to $t = \pi$. Also find the slope and second derivative at $t = \pi$.

2.36 Integrate $f(t)$ of Exercise 2.23 from $t = 0$ to $t = \pi$. Also find the slope and second derivative at $t = 0$.

Getting Help   Like other applications, Mathcad has built-in help . . . and then some. Click on menu item *Help* to get its help window. If you click on the home button, you'll get the Mathcad Resources window shown in Fig. 2.23, which has a tremendous amount of interactive information, including lots of reference material. Under Help you can also gain access to Web stuff, including interesting *User Forums*.

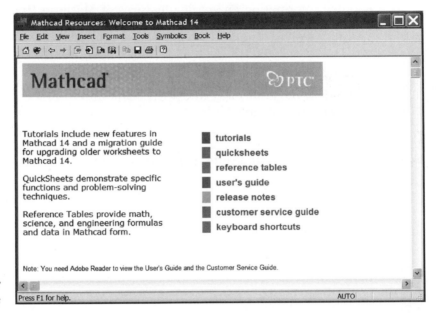

Figure 2.23:
Mathcad's Resources

Additional Exercises

Note: For some of these problems, you will find Mathcad's built-in Reference Tables very helpful. Click on menu item *Help . . . Reference Tables*.

Note: Be careful with variables and units in these problems! If you do several exercises in the same worksheet, there's a good chance you'll get confusing results (e.g., if you define V as velocity in one problem, you will *no longer* have V for volts as a built-in unit for a later problem). To avoid this, either use a separate worksheet for each exercise or make sure you are aware of this (and type, for example, **V:=volt** for the second problem)!

2.37 Find the mass (kg and lb) of the bicycle frame shown. The frame is made of tubular aluminum alloy (density $\rho = 2200$ kg/m^3), outside diameter D and inside diameter d.

2.38 Find the CG (center of gravity) of the bike frame shown. This is given by

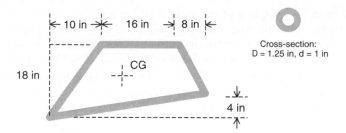

Cross-section:
D = 1.25 in, d = 1 in

10 in — 16 in — 8 in

18 in

CG

4 in

Figure E2.37, E2.38

$$x_{CG} = \frac{L_1 x_1 + L_2 x_2 + L_3 x_3 + L_4 x_4}{L_1 + L_2 + L_3 + L_4}$$

$$y_{CG} = \frac{L_1 y_1 + L_2 y_2 + L_3 y_3 + L_4 y_4}{L_1 + L_2 + L_3 + L_4}$$

where the L's, x's and y's are, respectively, the length and the x and y locations of the centers of the individual straight pieces.

2.39 A Venturi meter can be used to measure the flow of water in a pipe. As the flow converges at the throat, it speeds up and the pressure drops. By measuring the pressure drop Δp, the flow rate Q can be obtained from

$$Q = \sqrt{\frac{\dfrac{\Delta p}{\rho}}{\dfrac{1}{A_2^2} - \dfrac{1}{A_1^2}}}$$

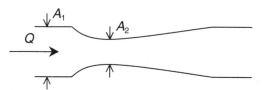

Figure E2.39

where ρ is the density of water and A_1 and A_2 are the pipe and throat area, respectively. If the pipe diameter is 6 in and the throat diameter is 4 in, find the flow rate in gal/min for $\Delta p = 3$ psi.

2.40 Normal blood pressure is "120 over 80." This means that when the heart pumps (systolic), the pressure is 120 mm Hg; when the heart relaxes (diastolic), the pressure is 80 mm Hg. Pressure must have units of force per area, but mm Hg means millimeters of mercury, a length! What's really happening is that your heart creates a pressure sufficient to support a column of mercury of 120 or 80 mm. Find the pressures in psi and kPa and as a percentage of atmospheric pressure. Note: Use the relation $p = \rho gh$, where p is pressure, $\rho = 13{,}300$ kg/m³ is the density of mercury, and h is the height of the column of mercury.

2.41 A jumping board at a swimming pool is $L = 5$ m long, $d = 40$ cm wide, and $b = 3$ cm thick. It is built from a composite material with a modulus of elasticity of $E = 1.25$ GPa. Find the deflection (in cm and in) of the end when a swimmer of 50 kg "weight" stands there. Note: For a cantilever beam, the end deflection is given by $\delta = \dfrac{MgL^3}{3EI}$, where M is the mass of the swimmer, and I is the second moment of area of the beam cross section, given by $I = \dfrac{bd^3}{12}$.

2.42 The height y of an object fired vertically is given by $y(t) = Vt - \frac{1}{2}gt^2$, where $V = 60$ mph. Find the height (ft) at $t = 0.5$ s, 1 s, and 2 s. Find the velocity at these same times. Note: Have Mathcad do the work! For the velocity define a new variable $v(t)$ to be equal to the derivative of $y(t)$.

2.43 A skydiver of mass $M = 60$ kg jumps from an airplane on two occasions. The first time she keeps her arms and legs in; the second time she is spread-eagled. For each jump, find her terminal speed V_t in m/s and mph, and her speed after 1 s and 10 s. Note: The terminal speed, $Vt = \sqrt{\dfrac{Mg}{k}}$,

where $k = 0.1$ kg/m (arms and legs in) and 0.3 kg/m (spread-eagled), is a coefficient of drag at the maximum speed, when the drag just equals the skydiver's weight. The speed at any time is given by

$$V(t) = V_t \tanh\left(\sqrt{\frac{g \cdot k}{M}}\, t\right)$$

2.44 The initial voltage V_0 across the capacitor shown is 100 V. Find the initial current (mA) in the circuit; and the current after 1 s. Find the total heat (J) generated in the resistor, and compare to the initial energy stored in the capacitor. Note: The equation for the current is $I(t) = \dfrac{V_0}{R} e^{-\frac{t}{R \cdot C}}$. Also,

the heat generated in the resistor is given by $Q = \displaystyle\int_{0 \cdot s}^{\infty \cdot s} R \cdot (I(t))^2 dt$ (the ∞

symbol can be found in the Calculus Toolbar), and the initial energy stored in the capacitor is $\frac{1}{2}V_0^2 C$.

2.45 A cylindrical storage tank of radius $R = 50$ cm and length $L = 3$ m is partially filled with a liquid to a depth $h = 70$ cm. Compute the volume (in liters and gallons) two ways:

(a) Using $V = 2L\displaystyle\int_0^h \sqrt{R^2 - (x - R)^2}\, dx$

(b) Using $V = LR^2(\pi - \theta + \frac{1}{2}\sin(2\theta))$
    where $\theta = \cos^{-1}\left(\dfrac{h - R}{R}\right)$.

$R = 25\ \text{k}\Omega$

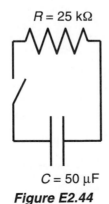

$C = 50\ \mu\text{F}$
Figure E2.44

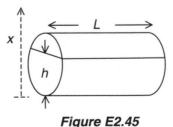

Figure E2.45

How to Numerically Solve Equations

Most problems in engineering and science involve mathematically describing the phenomenon using the appropriate physical laws. At some point we usually end up with an equation or equations to solve for the unknown or unknowns. Hopefully the unknown or unknowns can be made *explicit*. For example, we are familiar with the ideal gas equation

$$p\bar{v} = R_u T$$

where p is the pressure, $\bar{v}$ the molar volume, and T the temperature of the gas (R_u is the universal gas constant). We should have no problem finding, say, $\bar{v}$, if p and T are known (although Mathcad is useful even here because it will take care of the units for us). Often, an equation we wish to solve is *implicit* in the unknown. For example, instead of the ideal gas equation, a more accurate formula for modeling a gas is the Beattie-Bridgeman equation of state,

$$p\bar{v} = R_u T \left(1 - \frac{c}{\bar{v}T^3} \right)\left(1 + \frac{B_0}{\bar{v}}\left(1 - \frac{b}{\bar{v}} \right) \right) - \frac{A_0}{\bar{v}}\left(1 - \frac{a}{\bar{v}} \right)$$

where A_0, B_0, a, b, and c are known constants. This equation is explicit in p, but implicit and nonlinear in T and $\bar{v}$; it is not possible to find them in the form $T = \ldots$ or $\bar{v} = \ldots$.

More difficult problems involve *coupled* equations in which we must simultaneously solve several equations for several unknowns. For example, the equations that relate the pressure p, temperature T, density ρ, and speed V of an ideal gas (such as air) before ($_1$) and after ($_2$) a shock wave are

$$\rho_1 V_1 = \rho_2 V_2$$

$$p_1 + \rho_1 V_1^2 = \rho_2 + \rho_2 V_2^2$$

$$\frac{p_1}{\rho_1 T_1} = \frac{p_2}{\rho_2 T_2}$$

$$c_p T_1 + \tfrac{1}{2}V_1^2 = c_p T + \tfrac{1}{2}V_2^2$$

where c_p is the constant-pressure specific heat of the gas. These four equations can be used to find the pressure, temperature, density, and speed of the gas after the shock. They are tricky to solve because they are coupled, and also nonlinear in the unknowns.

In this chapter we will see how to use Mathcad to solve one or more equations for one or more unknowns. We'll do this using Mathcad's numerical methods. In Chapter 5 we'll look at Mathcad's analytic (symbolic) methods for solving the same kinds of problems.

3.1

Numerically Solving a Single Implicit Equation

Mathcad has three basic methods for solving for one unknown, as shown in Fig. 3.1. Note that Fig. 3.1 shows the function to be a cubic, but the methods (except polyroots) apply to basically any and all types of equation—algebraic, trigonometric, transcendental, and so on.

These methods are:

✓ Given . . . Find
✓ root
✓ polyroots (only used for polynomials)

We will now review all three by reproducing the worksheet of Fig. 3.1. First, although it's not strictly necessary to plot the equation, it's good practice because then you can often see where the solutions are located. We are interested in solving the cubic equation, or, alternatively, solving the equation $f(x) = 0$. You can go ahead and generate the worksheet through the line **Using "Given . . . Find":**. Note that we use the Boolean equals for the first equation (which is not actually used anywhere—it's just for us to look at); the definition of c has a squiggly line because we're redefining something Mathcad already knows (what does it use c for?), as discussed in Chapter 2; and we inserted and formatted an X-Y Plot. You can learn about graphing in Appendix: Graphing, but for now, if you want to create the graph:

1. In an empty region type **@**, or click on *Insert . . . Graph . . . X-Y Plot,* or click on the ⬚ icon in the Graph Toolbar.

2. Click on each of the three black placeholders on the horizontal axis and type (starting with the one on the left), **−2, x,** and **3**.

3. Click on the middle black placeholder of the vertical axis and type **f(x)**.

4. Press **Enter** to exit the graph.

5. You can move the graph by clicking once on it and moving the mouse to the edge, where you'll get a small black hand, ✋ ; use this to move the graph.

6. You can resize the graph by clicking once on it and then dragging on the small placeholder, ▪ , on the right, bottom, or corner boundary of the graph.

7. If you really want to get into graphing, open up the Graph Toolbar, click once on your graph, and then play with the *Zoom* (🔍) and *Trace* (⚘) icons. The *Zoom* icon lets you select a small region of the graph to zoom in

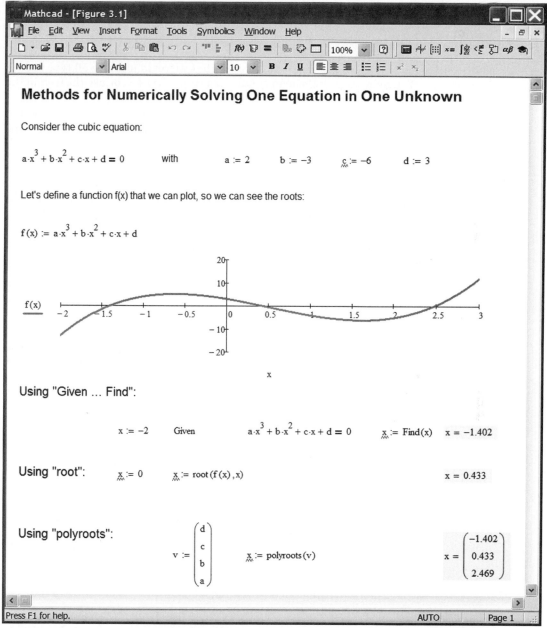

Methods for Numerically Solving One Equation in One Unknown

Consider the cubic equation:

$a \cdot x^3 + b \cdot x^2 + c \cdot x + d = 0$ with $a := 2$ $b := -3$ $c := -6$ $d := 3$

Let's define a function f(x) that we can plot, so we can see the roots:

$f(x) := a \cdot x^3 + b \cdot x^2 + c \cdot x + d$

Using "Given ... Find":

$x := -2$ Given $a \cdot x^3 + b \cdot x^2 + c \cdot x + d = 0$ $x := Find(x)$ $x = -1.402$

Using "root": $x := 0$ $x := root(f(x), x)$ $x = 0.433$

Using "polyroots":

$$v := \begin{pmatrix} d \\ c \\ b \\ a \end{pmatrix}$$ $x := polyroots(v)$ $$x = \begin{pmatrix} -1.402 \\ 0.433 \\ 2.469 \end{pmatrix}$$

on (e.g., around $x = -1.5$), and then the *Trace* icon lets you read out graph values as you use the arrow keys (← and →).

Using Given . . . Find to Numerically Solve an Equation

This is the most useful and widely used method. The required steps are:

1. You *must* provide an initial guess for the unknown (in Fig. 3.1 we set $x = -2$, based on the graph. However, you could choose another value such as 0 or 1).

2. Type the math region **Given**. This is a built-in math function in Mathcad. Do not use the **Spacebar** at any point or you'll convert it to a text region!

3. To the right of or below Given, type the equation you wish to solve. In Fig. 3.1 we typed the cubic equation, but we could just as well type **f(x)=0.** You must use the Boolean equals sign (type **Ctrl + =**, or use the icon in the Boolean Toolbar).

4. Use **Find( )** to obtain the solution. Inside the parentheses, type the variable you're solving for (in our case type **x**). Note that if you type **Find(x)=** without assigning the answer to x or to any other variable, you'll just get the numerical answer (try it first, then delete it); if you want x to *be* the solution, type **x:Find(x)** as we have in Fig. 3.1.

Try these steps to get the answer shown in Fig. 3.1 (to highlight the answer, right-click on it and choose *Properties . . . Display . . . Highlight Region*).

These steps are very straightforward, but there are some subtleties to be aware of:

1. Because this is a numerical method, the solution you get depends on your initial guess value. The solution obtained will usually, but not always, be the one closest to your initial guess. To find additional solutions, change the initial guess value (e.g., try $x = 1$ in the worksheet of Fig. 3.1).

2. Again, because this is a numerical method, sometimes Mathcad fails to find a solution. In this case try varying the initial guess or changing its type (see the discussion of common mistakes in the next section), or setting up the problem differently, although success is not guaranteed!

3. If you are solving for one unknown, you can obviously only have one equation to solve; however, you can use additional constraint equations (e.g., try rearranging the equations so you can include **x > 0** between the Given and Find expressions in the worksheet of Fig. 3.1).

4. You can use units with Given . . . Find (see Example 3.1).

5. You can change the accuracy with which Given . . . Find finds a solution. Click on *Tools . . . Worksheet Options* to get a window where you can click on the Built-In Variables tab and change the value of TOL. TOL is a variable that Mathcad uses to determine when a sufficient accuracy has been achieved—a smaller TOL value yields a more accurate result (at the expense of taking longer to find).

6. Instead of using Given . . . Find, you can use Given . . . Minimize or Given . . . Maximize (see Section 3.3).

We will look at several examples of using Given . . . Find, but first let's mention some common mistakes in its use.

Common Mistakes in Using Given . . . Find

Figure 3.2 shows examples of common errors with using Given . . . Find for numerically solving an equation (you may wish to reproduce this worksheet as an exercise). These mistakes are:

1. Not providing an initial guess.

2. Not using the symbolic equals between Given and Find.

3. Not using the computed result. This is not a mistake if all you want to do is find the unknown value (in Fig. 3.2 the answer is 0.433), but if you wish to use the answer for a later calculation you should remember to set *x* (or any other variable you wish) equal to the Find.

4. Using the incorrect guess type. This is a subtle error: if you expect the solution to be complex, you must give a complex (or imaginary) initial guess. To type a complex number you create the "i" by typing **i** next to a number (e.g., to get 3 + 5i type **3+5i**). Notes: If the imaginary part is 1, you *must* type a 1 next to the i (e.g., to get 2 + i type **2+1i**); You can change the default *i* symbol to *j* by using *Format . . . Result . . . Display Options*. Try changing the initial guess in Fig. 3.2 to *i*.

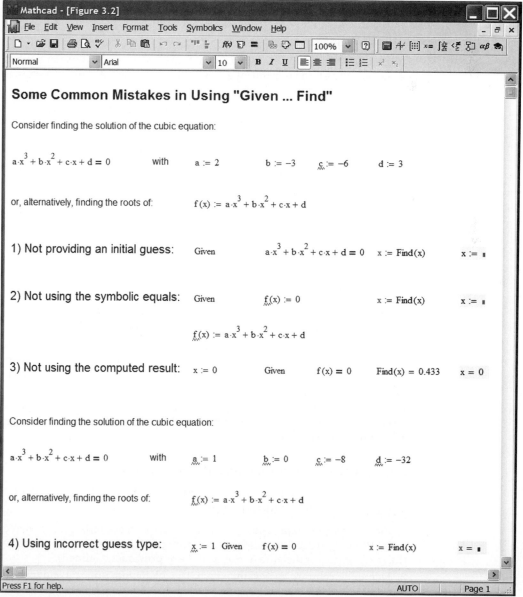

Figure 3.2: Common mistakes in using Given . . . Find

Using root to Numerically Solve an Equation

To find the root of a single expression, you must *either*:

1. Provide an initial guess for the unknown (in Fig. 3.1 we set $x = 0$), and use the root function with two arguments—the function whose root you wish to find and the variable of interest (in Fig. 3.1 we typed **x:root(f(x),x)**) *or*

2. Include two extra arguments (instead of providing an initial guess), telling Mathcad x values between which you know there is a root (in Fig. 3.1 we could have typed **x:root(f(x),x,−2,−1)** instead of setting $x = 0$ and typing **x:root(f(x),x)**).

Try these steps to get the answer shown in Fig. 3.1.

As with Given . . . Find, these steps are straightforward, but again there are some subtleties that you should be aware of:

1. Because this is a numerical method, the solution you get depends on your initial guess value. The solution obtained will usually, but not always, be the one closest to your initial guess. To find additional solutions, change the initial guess value.

2. Sometimes Mathcad fails to find a solution. In this case try varying the initial guess or changing its type (e.g., complex), or setting up the problem differently, although success is not guaranteed!

3. You can use units with root (see Example 3.2).

4. You can change the accuracy with which root finds a solution. Click on *Tools . . . Worksheet Options* to get a window where you can click on the Built-In Variables tab and change the value of TOL. A smaller TOL value yields a more accurate result (at the expense of taking longer to find).

Common Mistakes in Using root

Figure 3.3 shows some of the common errors in using root. These mistakes are:

1. Using the wrong case—you must type **root**, not **Root**.

2. Not providing an initial guess (when using **root(f(x),x)**)

3. Not using the computed result. This is not a mistake if all you want to do is find the unknown value, but if you wish to use the answer for a later calculation, you must remember to set x (or any other variable you wish) equal to the root function.

4. Using the incorrect guess type. As with Given . . . Find, this is a subtle error: if you expect the solution to be complex, you must give a complex (or imaginary) initial guess.

Using polyroots to Numerically Solve an Equation

This is a powerful but limited method—it can only find the roots of a polynomial. The required steps are:

1. Make sure the expression you're finding the roots of is in the correct form. It must be a polynomial (e.g., as in Fig. 3.1, $ax^3 + bx^2 + cx + d$, where the polynomial coefficients $a, b, c,$ and d are previously defined).

2. Define a column vector in which the elements are the polynomial coefficients in order, with the first coefficient being the constant (the

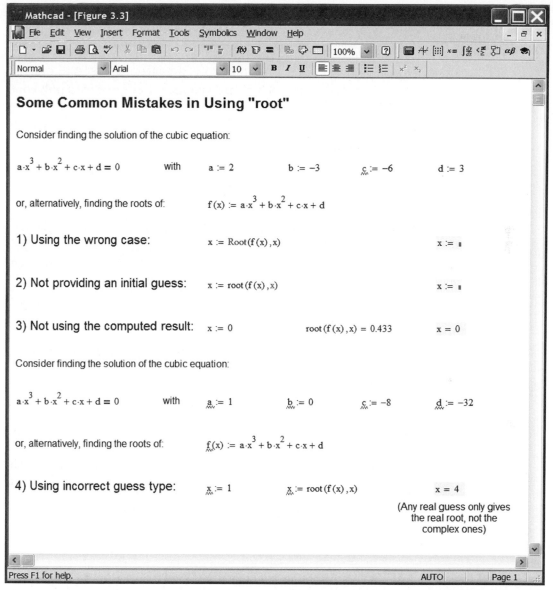

Some Common Mistakes in Using "root"

Consider finding the solution of the cubic equation:

$$a \cdot x^3 + b \cdot x^2 + c \cdot x + d = 0 \qquad \text{with} \qquad a := 2 \qquad b := -3 \qquad c := -6 \qquad d := 3$$

or, alternatively, finding the roots of:
$$f(x) := a \cdot x^3 + b \cdot x^2 + c \cdot x + d$$

1) Using the wrong case: $\qquad x := \text{Root}(f(x), x) \qquad\qquad\qquad x := \blacksquare$

2) Not providing an initial guess: $\quad x := \text{root}(f(x), x) \qquad\qquad\qquad x := \blacksquare$

3) Not using the computed result: $\quad x := 0 \qquad\qquad \text{root}(f(x), x) = 0.433 \qquad x = 0$

Consider finding the solution of the cubic equation:

$$a \cdot x^3 + b \cdot x^2 + c \cdot x + d = 0 \qquad \text{with} \qquad a := 1 \qquad b := 0 \qquad c := -8 \qquad d := -32$$

or, alternatively, finding the roots of:
$$f(x) := a \cdot x^3 + b \cdot x^2 + c \cdot x + d$$

4) Using incorrect guess type: $\qquad x := 1 \qquad\qquad x := \text{root}(f(x), x) \qquad\qquad x = 4$

(Any real guess only gives
the real root, not the
complex ones)

Press F1 for help.   AUTO   Page 1

**Figure 3.3:
Common mistakes
in using root**

coefficient for x^0; in Fig. 3.1 this is d). To create a column vector you can either type **Ctrl + M** (M for matrix) or use the Vector and Matrix Toolbar and click on the Matrix or Vector icon. In either case you'll get an Insert Matrix window waiting for you to enter the number of rows and columns (in Fig. 1.3 we created a vector **v** as a 4×1 matrix). Then simply type in the matrix placeholders the coefficients (or their numerical values); to go from placeholder to placeholder either click with the mouse or press **Tab**.

3. Use the polyroots expression with the only argument being the vector created in Step 2 (in Fig. 3.1 we typed **x:polyroots(v)**).

Try these steps to get the answer shown in Fig. 3.1. Because this is a numerical method, there is no guarantee of a solution or of its accuracy. In this case you can try asking Mathcad to use an alternative algorithm for the polyroots function by right-clicking on it and choosing the Companion Matrix rather than the Laguerre default algorithm (don't ask!)

Probably the most common mistake is typing the function incorrectly: make sure you type **polyroots**, not **Polyroots** or **polyroot**!

We will now do several complete examples of Given . . . Find, roots, and polyroots; reproducing the worksheets will be good practice in using these features. Each example illustrates additional features (some bells and whistles) of these functions.

Example 3.1: Using Given . . . Find with Units: Free Vibration of a Mass—Beam As a simulation of the vibration of part of a building, a mass $M = 100$ kg is attached to an iron beam of length $L = 15$ ft and diameter $d = 3$ in. Find the natural frequency of the system two ways: (a) a simple method, in which we assume the beam is a simple spring (i.e., ignore the beam mass); (b) the exact method, in which we allow for the fact the beam itself can vibrate.

The completed worksheet for this problem is shown below. The stiffness of a beam is given by $k = \dfrac{\pi d^2}{4} \dfrac{E}{L}$, where E is modulus of elasticity. The natural frequency of a spring-mass system (ignoring the beam mass) is $\omega = \sqrt{\dfrac{k}{M}}$. This equation is easy to solve for the frequency ω. On the other hand, the equation for the frequency of vibration ω of a mass and beam allowing for the beam mass is

$$\omega L \sqrt{\frac{\rho}{E}} \, \tan\!\left(\omega L \sqrt{\frac{\rho}{E}} \right) = \frac{M_{\mathrm{b}}}{M}$$

where ρ is the density of steel and M_{b} is the beam mass. Solving for ω is tricky without Given . . . Find!

You can go ahead and type in a new worksheet everything as shown in Fig. 3.4. The only points of concern are:

1. Note that the given problem is given in mixed units, which Mathcad, and specifically Given . . . Find, can handle; the default units are SI, so for our own interest, we evaluated some variables and then clicked on the result and inserted different units (e.g., we evaluated the mass M in lb).

2. Remember that for Greek letters you can either use the Greek Symbol Toolbar, or instead type the equivalent Roman letter and immediately press **Ctrl + G**.

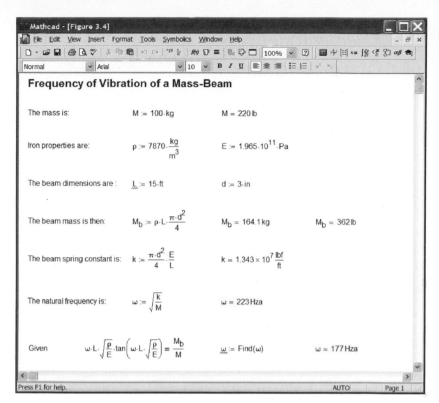

Figure 3.4:
The completed
worksheet for
Example 3.1

3. The density and modulus of elasticity can be obtained from any good solid mechanics text or online. (Mathcad's own Reference Tables, found under Help, are very useful!)

4. The beam-mass notation, M_b, has a subscript. To get a *literal subscript* (as opposed to a math subscript, used when we define matrices or vectors —see Chapter 4), you *must* use a period; for example, type **M.b** to get $\boxed{M_b}$. (The period in the subscript disappears when you exit the expression.)

5. For the two frequencies, the result is first displayed with units of $\frac{1}{s}$; we must double-click on the units placeholder in the answer to get the Insert Unit window, select *Frequency,* and choose *Hza* (Mathcad's unit for cycles/s).

Note that the natural frequency obtained when we ignore the beam mass is higher than the frequency obtained when we include the beam mass, which intuitively makes sense: the "heavier" the mass, the slower it vibrates. Actually, there are an infinite number of solutions to this problem, corresponding to modes of vibration in which different regions of the beam vibrate in step or out of step with the mass. You can try to find a few more as an exercise.

Example 3.2: Using root with Units: Free Vibration of a Mass-Beam Repeat Example 3.1 using the root function.

In a new worksheet we can reproduce everything in Fig. 3.4 except we delete the Given . . . Find regions and instead use root, as shown in Fig. 3.5. Note that we used the natural frequency value as our required initial guess for the root function.

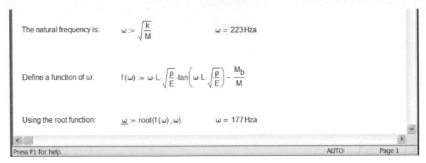

*Figure 3.5 :
The last three lines
for Example 3.2*

Example 3.3: Using Given . . . Find and polyroots with Units: Top Speed of a Car

A car's engine produces at most $P = 95$ hp at the wheels. The resistance to motion is due to rolling resistance in the tires and bearings, $F_{Rolling} = aV$ where $a = 5$ N/(m/s) and V is the speed, and aerodynamic drag, $F_{Aero} = bV^2$ where $b = 0.4$ N/(m/s)^2. Find the top speed of the car.

For this problem, the top speed will be when the power produced by the engine is entirely consumed by work done per time against the total drag (recall work is force times distance, so rate of work is force times velocity) or

$$P = (F_{Rolling} + F_{Aero})V = (aV = bV^2)V$$

We can use Mathcad's Given . . . Find to solve this equation. In a new worksheet you can reproduce everything in Fig. 3.6. This indicates again how easy and powerful Given . . . Find is. Remember, the Given is a math region, not a text region!

We can also solve the above equation using polyroots because it is a polynomial:

$$bV^3 + aV^2 - P = 0$$

The solution is shown in Fig. 3.7. There are a few things to point out:

1. Polyroots doesn't handle units, so we need to redefine a, b, and P as dimensionless quantities, and we (not Mathcad) have to worry about units. A benefit of SI units is that if we use basic units (N, kg, m, s), we don't need to worry too much. Hence we ask Mathcad to evaluate the 95 hp in SI units; then, we redefine P without units, as shown.

2. We need to be careful in defining the polynomial; we typed in the non-computing polynomial equation so it's clear what the equation is.

3. We use the steps for polyroots discussed earlier in this chapter: define the column vector consisting of the polynomial coefficients in the correct order; use the polyroots solver; interpret results.

4. Note that we get three solutions to the polynomial, as we should, because it's a cubic equation. We know from physical reasoning that there must

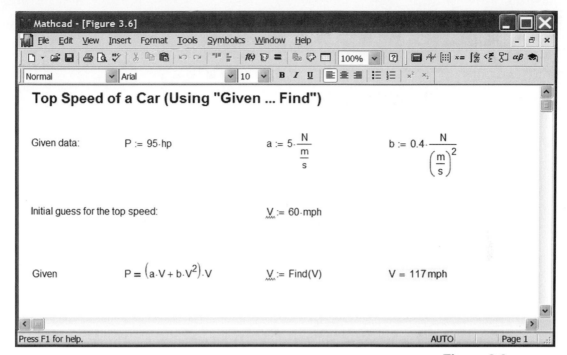

Top Speed of a Car (Using "Given ... Find")

Given data: $P := 95 \cdot hp$ $a := 5 \cdot \dfrac{N}{\dfrac{m}{s}}$ $b := 0.4 \cdot \dfrac{N}{\left(\dfrac{m}{s}\right)^2}$

Initial guess for the top speed: $V := 60 \cdot mph$

Given $P = \left(a \cdot V + b \cdot V^2\right) \cdot V$ $V := Find(V)$ $V = 117\,mph$

Figure 3.6: The Given . . . Find worksheet for Example 3.3

only be one top speed, so the other results must be physically nonsensical, and they are—the other two solutions are complex.

5. To get our solution in mph, we must realize that the solution obtained has no dimensions, so we must manually redefine the answer and insert the SI units; finally, we can evaluate what we just defined, and insert the desired units.

It is pretty clear that the polyroots method is not very useful: it only solves polynomials, and even then only without units!

Example 3.4: Using "Given . . . Find" with Calculus: Depletion of Oxygen in a River The wastewater from a power station enters a river, and as it flows downstream, there is a depletion of oxygen concentration in the river. The concentration c (%) at any downstream point x (m) is given by

$$c(x) = A - B\left(e^{-\frac{x}{L}} - e^{-\frac{5x}{L}}\right)$$

where $A = 5\%$, $B = 4\%$, and $L = 3$ km. Find the minimum concentration of oxygen, and the location downstream (m) of the power station at which it occurs.

For this problem, we can find the location of the maximum or minimum by setting the first derivative equal to zero:

$$\frac{dc(x)}{dx} = 0$$

41

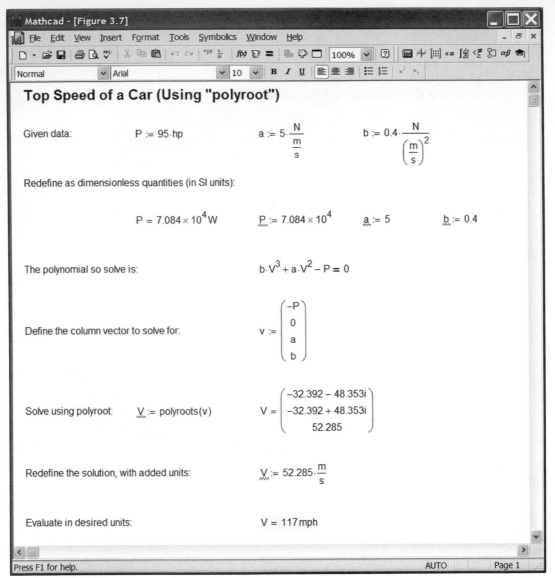

Mathcad - [Figure 3.7]

File Edit View Insert Format Tools Symbolics Window Help

Normal   Arial   10   **B** *I* U

Top Speed of a Car (Using "polyroot")

Given data: $P := 95 \cdot hp$ $a := 5 \cdot \dfrac{N}{\dfrac{m}{s}}$ $b := 0.4 \cdot \dfrac{N}{\left(\dfrac{m}{s}\right)^2}$

Redefine as dimensionless quantities (in SI units):

$$P = 7.084 \times 10^4 \, W \qquad P := 7.084 \times 10^4 \qquad a := 5 \qquad b := 0.4$$

The polynomial so solve is: $b \cdot V^3 + a \cdot V^2 - P = 0$

Define the column vector to solve for: $v := \begin{pmatrix} -P \\ 0 \\ a \\ b \end{pmatrix}$

Solve using polyroot: $V := polyroots(v)$ $V = \begin{pmatrix} -32.392 - 48.353i \\ -32.392 + 48.353i \\ 52.285 \end{pmatrix}$

Redefine the solution, with added units: $V := 52.285 \cdot \dfrac{m}{s}$

Evaluate in desired units: $V = 117 \, mph$

Press F1 for help.   AUTO   Page 1

Figure 3.7:
The polyroots
worksheet for
Example 3.3

Hence, our goal is to solve this equation with $c(x)$ defined by the previous equation. You can go ahead and create the worksheet shown in Fig. 3.8, with the following points:

1. In defining A and B we use % just like any other unit.
2. To get the exponential function, just type **e**!
3. In this example, it's useful to plot the graph. To do so:
   a. In an empty region type **@**, or click on *Insert . . . Graph . . . X-Y Plot*, or click on the ☒ icon in the Graph Toolbar.
   b. Click on each of the three black placeholders on the horizontal axis and type (starting with the one on the left), **0**, **x**, and **5000*m**.

42

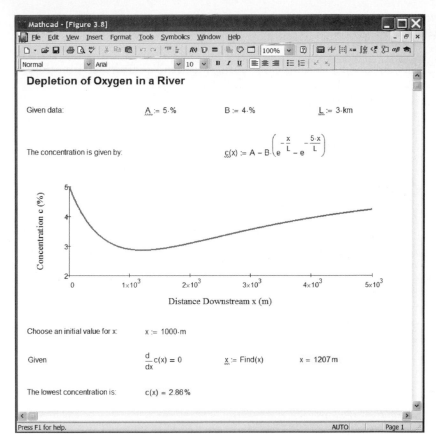

**Figure 3.8:
The completed
worksheet for
Example 3.4**

The worksheet content shown in the figure:

Depletion of Oxygen in a River

Given data: $A := 5 \cdot \%$ $B := 4 \cdot \%$ $L := 3 \cdot km$

The concentration is given by: $c(x) := A - B \cdot \left(e^{-\frac{x}{L}} - e^{-\frac{5 \cdot x}{L}} \right)$

Choose an initial value for x: $x := 1000 \cdot m$

Given $\dfrac{d}{dx} c(x) = 0$ $x := Find(x)$ $x = 1207 \, m$

The lowest concentration is: $c(x) = 2.86 \%$

c. Click on the middle black placeholder of the vertical axis and type **c(x)/%** (we divide by % otherwise the numbers on the vertical axis will be decimal percent, e.g., 5% would display as 0.05).

d. Double-click on the graph to get a Formatting Currently Selected X-Y Plot window and then change the axes style to crossed, increase the line weight, hide arguments, and add x- and y-axis labels.

e. Press **Enter** to exit the graph.

f. Move the graph by clicking once on it and moving the mouse to the edge, where you'll get a small black hand, 🖐 ; use this to move the graph.

g. Resize the graph by clicking once on it and then dragging on the small placeholder, ⊣ , on the right, bottom, or corner boundary of the graph.

4. The graph helps us choose a good guess value for the location of the minimum.

5. Set up the Given . . . Find to find the x at which the slope is zero. Do not try to type it to obtain the derivative! Instead you must either open the Calculus Toolbar and select the Derivative icon, or use accelerator keys **Shift + /**. Note also you must type **c(x)** not just **c** in the derivative. We will have much more to say about calculus in Chapter 5.

6. Once we obtain the x at which c is a minimum, we can easily evaluate this value of c by computing c(x).

Mathcad numerically solves the single equation for x!

Note that the equation is *not* a differential equation, but merely a constraint equation for *x*. We will use Mathcad to solve differential equations in Chapter 6. At the end of this chapter, we will also discuss solving Example 3.4 using Mathcad's minimize function.

Exercises

Note: For some of these problems, you will find Mathcad's built-in Reference Tables very helpful. Click *Help . . . Reference Tables!*

Note: Be careful with variables and units in these problems! If you do several exercises in the same worksheet, there's a good chance you'll get confusing results (e.g., if you define *V* as velocity in one problem, you will *no longer* have *V* for volts as a built-in unit for a later problem). To avoid this, either use a separate worksheet for each exercise, or make sure you are aware of this and type, for example, **V:=volt** for the second problem!

Cross-section:

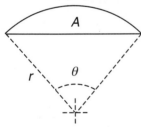

b ←--┆*a*

Figure E3.1

3.1 An "O" ring (a torus) is to be designed for a car suspension seal. It must have an inner radius of *a* = 2 cm and have a volume of 0.65 cm3. Find the outer radius *b* (cm), and then the surface area *S* (cm2).

3.2 Find the angle θ (°) shown so that the lawn area contained is *A* = 15 m² when the radius *r* = 30 m. Then find the total length of the lawn perimeter.

3.3 Find the angle θ (°) shown so that the centroid of the "pizza slice" is x_C = 7.5 in when the radius *r* = 12 in.

3.4 The height *y* of an object fired vertically is given by $y(t) = Vt - \frac{1}{2}gt^2$, where *V* = 60 mph. Find the two times *t* (s) at which the height is *h* = 100 ft and is *h* = 120 ft.

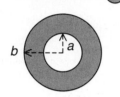

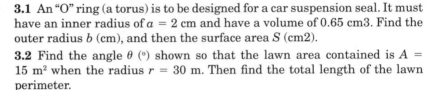

Figure E3.2

3.5 For Exercise 3.4, find the maximum height by solving for the time *t* at which $\dfrac{dy(t)}{dt}$ = 0m/s, then evaluating *y(t)*. Compare to the known result $h = \dfrac{V^2}{2g}$.

3.6 We need to find the pressure drop, given by $\Delta p = \rho f \dfrac{L}{D} \dfrac{V^2}{2}$, in *L* = 1 mi (miles) of water pipe, with pipe diameter *D* = 1 in. The velocity is given by $V = \dfrac{4Q}{\pi D^2}$, where *Q* is the flow rate in the pipe and ρ is the density of water. The term *f* is the friction factor, used in calculating the amount of friction. For turbulent flow we can obtain the friction factor from

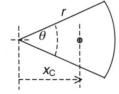

r

θ

x_C

Figure E3.3

$$\frac{1}{\sqrt{f}} = -2\log\left(\frac{e/D}{3.7} + \frac{2.51}{Re\sqrt{f}}\right)$$

where *Re* is the Reynolds number and *e* = 0.01 in is the roughness of the pipe surface. The Reynolds number in turn is given by $Re = \dfrac{\rho VD}{\mu}$ where μ is the fluid viscosity.

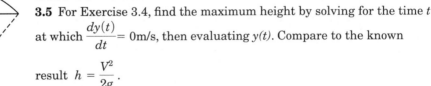

Note that f and Re end up being dimensionless. Find the friction factor and pressure drop (psi) for a flow rate of (a) 1 gal/min and (b) 4 gal/min. Use Given . . . Find.

3.7 Repeat Exercise 3.6 using root.

3.8 Use root to solve $ln(x + 1) = \left(\dfrac{x}{4}\right)^2$ for $x > 0$.

3.9 Use root to find the three positive solutions of $sin(x) = \dfrac{x}{3\pi}$.

3.10 Use root to find the four solutions of $tan(2x) = cos(x)$ between 0 and 2π.

3.11 Use root to solve $x^3 + 3x^2 - 4x - 12 = 0$.

3.12 Use root to solve $x^4 - x^3 - 7x^2 + x + 6 = 0$.

3.13 Use polyroots to solve Exercise 3.11.

3.14 Use polyroots to solve Exercise 3.12.

3.2

Numerically
Solving
Simultaneous
Equations Using
Given . . . Find

Mathcad's Given . . . Find method, as we have seen, is an excellent tool for solving a single implicit equation. The method really comes into its own when it is used to solve a set of equations. The equations to be solved could be linear or nonlinear, and involve integrals or differentials. If the equations are linear, linear algebra methods can be used instead of the Given . . . Find method (see Chapter 4 for details on this).

The procedure for using the Given . . . Find method is as described in Section 3.1, with the following additional comments:

1. Each of the unknowns must be given an initial guess value (based on common sense, engineering experience, or examination of the engineering problem by, for example, graphing using Mathcad).

2. The set of equations to be satisfied must be placed between the Given and Find statements. Care must be taken that the equations are encountered by Mathcad *after* the Given and *before* the Find. Note that the Boolean equals *must* be used here!

3. The solution is given as a vector of values. If the unknowns have differing units (e.g., if one unknown was time, another pressure, and a third velocity), we have to be a little careful in using the method because normally a vector must have the same units for all elements (see Example 3.6).

4. As with single equation problems, instead of the above output, we can define the solution as new quantities for use in a later equation.

5. You can sometimes force Mathcad to search for a particular solution by adding additional constraint equations.

We will do several examples of simultaneous equations, but first let's list some common errors.

Some Typical Errors Using Given . . . Find

1. There are an insufficient number of equations. For each unknown, one independent equation must be given. If you did not provide enough equations, check the engineering problem for conceptual errors.

2. The unknowns have differing units (see Example 3.6 for fixing this).

3. The computed results are not used. This is not a mistake if all you want to do is find the unknown values, but if you wish to use the answers for later calculations, you should set a column vector equal to the Find.

4. The initial guesses do not lead to a solution, or they lead to an unwanted solution. Because this is a numerical method, occasionally the method will fail, even if the problem is "well-posed." If this occurs, the only options are to try different initial conditions or to add an additional constraint equation (see Example 3.7).

Example 3.5: Using "Given . . . Find" for Simultaneous Equations: Currents in a DC Network Given the network shown in Fig. 3.9, find the currents I_1, I_2, and I_3.

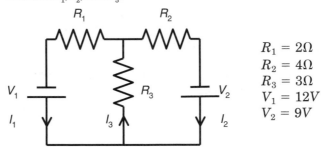

$R_1 = 2\Omega$
$R_2 = 4\Omega$
$R_3 = 3\Omega$
$V_1 = 12V$
$V_2 = 9V$

Figure 3.9:
Currents in a
DC circuit

Applying Kirchoff's law and summing voltage drops across each battery, we obtain three equations for the three unknowns, I_1, I_2, and I_3

$$I_1 + I_2 - I_3 = 0$$
$$R_1 I_1 + R_3 I_3 = V_1$$
$$R_2 I_2 + R_3 I_3 = V_2$$

The worksheet of Fig. 3.10 shows the solution; it's a good idea to try and reproduce it.

We can make some comments here:

1. All of the subscripts are literal subscripts. To get a literal subscript you must use a period; for example, type **R.1:2*ohm** to get $R_1:=1.ohm$. In Chapter 4 we will see that there are also *math subscripts*, designed for use with vectors and matrices.

2. The equations for this problem are linear, so we could have used linear algebra methods to solve them; we will do so in Chapter 4.

3. We will explore Mathcad's vector and matrix math in Chapter 4, but for now, if we wish to use the results obtained, we must use the following alternate approach for the Find:

    a. Create a vector containing the unknowns (recall that to create a column vector you can either type **Ctrl + M** (M for matrix) or use the Vector and

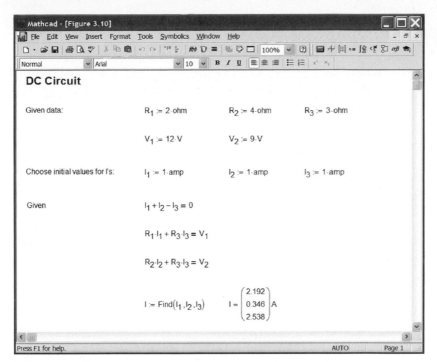

Figure 3.10:
The completed
worksheet for
Example 3.5

Matrix Toolbar and click on the Matrix or Vector icon. In either case you'll get an Insert Matrix window waiting for you to enter the number of rows and columns (in Fig. 3.11 we created a 3×1 matrix).

b. Until we get to Chapter 4 we will not be using math subscripts, so use literal subscripts as shown.

c. The values of I_1, I_2, and I_3 are now available for use in calculations as needed. As an example, we compute the power consumption in the system.

Example 3.6: Using "Given . . . Find" for Simultaneous Equations with Units: Car Accident After a hit and run fatal accident, forensic engineers figure out from the distance the man was thrown that the impact gave the body an initial speed of 45 mph. The tire skid marks indicate the car speed immediately after collision was about 25 mph. The man weighed 220 lb, and the engineers estimate from body damage that the collision

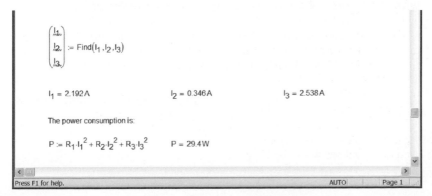

Figure 3.11:
Alternative use of Find
for Example 3.5

must have absorbed about 4500 J of energy. Estimate the car's mass and initial speed.

For this problem, we use conservation of momentum and energy:

$$m_C v_C = m_C v'_C + m_B v'_B$$

$$\tfrac{1}{2} m c v_C^2 = \tfrac{1}{2} m c v'^2_C + \tfrac{1}{2} m_B v'^2_B + \Delta E$$

where m_C and m_B are the car and body mass, v_C is the car speed before the crash, v'_C and v'_B are the car and body speeds after the crash, and ΔE is the energy absorbed in the impact. The unknowns are m_C and v_C. These two equations form a *nonlinear coupled* set for the two unknowns (they can be solved analytically, but here we'll use Mathcad's Given . . . Find anyway).

The completed worksheet is shown in Fig. 3.12. It looks like the car was small (under 2000 lb) and was not speeding (speed before collision was about 30 mph). The only tricky parts are:

1. Be careful in using lb—it is a unit of mass; if you intend to use pounds force, you must type lbf!

2. Remember that for Greek letters you can either use the Greek Symbol Toolbar or type the equivalent Roman letter and immediately press **Ctrl + G** (e.g., for Δ, type **D** then immediately press **Ctrl + G**)

3. Once again we use *literal subscripts*.

4. To get the primes, type ` (on most keyboards, the key that also has tilde (~)).

5. Because the two unknowns have differing units (i.e., mass and velocity), you cannot define the answer to be a column vector! For example, see Fig. 3.13 for what happens if you try to do so. You get an odd-looking error message, which is trying to say "Solution has units of mass based on the first term, but the second term has length · time $^{-1}$, which is illegal." *This is one of the most common errors when solving for several unknowns that have differing units!*

Example 3.7: Using "Given . . . Find" for Simultaneous Equations with Constraint Equations: Length of a Connector An oval piece of tubing (part of a lawn chair design) 0.80 m by 0.50 m is to have a strengthening member as shown in Fig. 3.14. The constraints are that the member must be at $\theta = 30°$ to the horizontal, and at a height 0.15 m above the oval center. Find the length of the member.

For this problem we need the equations for an ellipse and straight line. Using the notation defined by Fig. 3.14, we have

$$\left(\frac{x}{a}\right)^2 + \left(\frac{y}{b}\right)^2 = 1$$
$$y = x\tan(\theta) + c$$

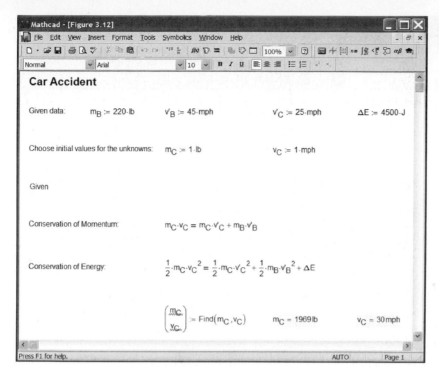

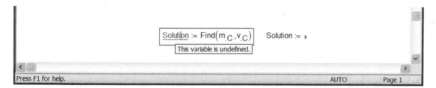

Figure 3.12:
The completed
worksheet for
Example 3.6

Figure 3.13:
Trying to solve for
unknowns with
differing units, using
a column vector

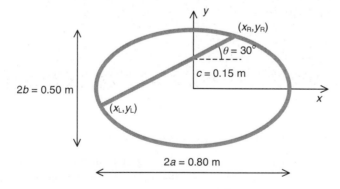

Figure 3.14:
Geometry of chair
assembly

We need to solve these *nonlinear coupled* simultaneous equations for x and y to find the two points (x_L, y_L) and (x_R, y_R) at which the line and ellipse meet. Then, to find the length L of the connector, we use

$$L = \sqrt{(x_L - x_R)^2 + (y_L - y_R)^2}$$

The completed worksheet for this problem is shown in Fig. 3.15.

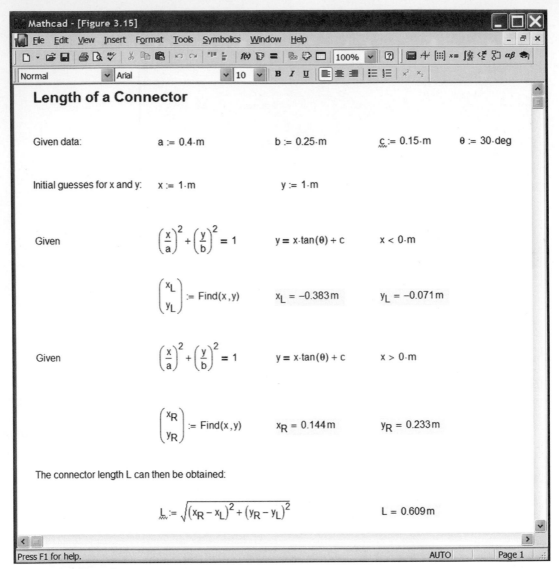

Figure 3.15:
The completed
worksheet for
Example 3.7

We can make some more comments here:

1. We defined the angle θ in degrees. Mathcad computes trigonometric functions in radians, but you can still define angles in degrees; Mathcad takes care of the nitty-gritty!

2. We used the Given . . . Find twice because we wanted both solutions to the set of equations. We used the same initial guess values both times and *forced* Mathcad to find two different solutions by using constraint equations within the solve blocks; in the first block we typed **x<0*m** and in the second **x>0*m**. *Constraints are very useful for forcing Mathcad to find the solution you want out of several possible solutions.*

3. As with all the worksheets we have created, they are live, so you can do interesting what-ifs by changing given data.

Example 3.8: Using Given . . . Find for Simultaneous Equations with Functions: Location of Support Beams As part of a modern roof design for a new museum, shown in Fig. 3.16, we need to locate two points at which to place vertical support beams. The beams must be placed under points on the surface at which the two different curves intersect, and also with the construction limitation that $x = 2y$ (x and y are non-dimensional ground coordinates). Find the two (x, y) locations. The two surfaces are given by

$$g(x, y) = 1 - \left(\frac{x}{4}\right)^2 - \left(\frac{y}{4}\right)^2$$

$$f(x, y) = 1 + J_0(6x)e^{-\frac{y}{2}}$$

This problem sounds a bit complicated, but it boils down to solving for x and y for $g(x, y) = f(x, y)$ and $x = 2y$.

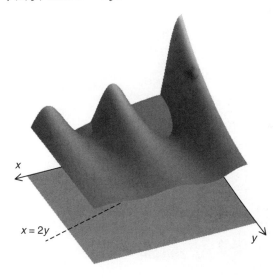

**Figure 3.16:
Shape of museum
roof**

The completed worksheet is shown in Fig. 3.17. You should try to reproduce it. The only tricky parts are:

1. The function $f(x, y)$ includes one of Mathcad's many built-in funtions, the zeroth-order Bessel function of the first kind, $J_0(z)$; to use it simply type **J0(z)** (in our case **J0(6·x)**).

2. The graph takes a bit of work (graphing is covered in detail in Appendix: Graphing). To create it:

    a. In an empty region type **Ctrl + 2**, or click on *Insert . . . Graph . . . Surface Plot,* or click on the ⬛ icon in the Graph Toolbar.

    b. At the black placeholder type **f,g** and press **Enter.** You'll get a pretty ugly wire-frame graph!

    c. Move and resize to a convenient place and size graph by clicking once on it and moving the mouse to the edge of the entire graph object to get the small black hand, ✋ , or to the edge or corner of the graph itself, to

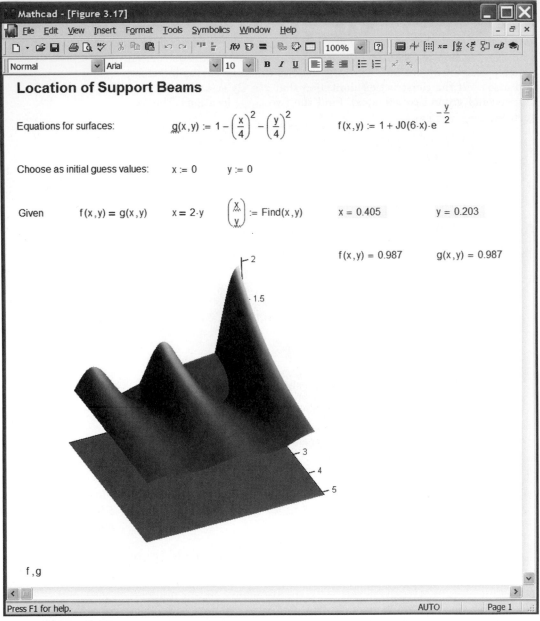

The image shows a Mathcad worksheet window titled "Mathcad - [Figure 3.17]" with menu bar and toolbars.

Location of Support Beams

Equations for surfaces:

$$g(x,y) := 1 - \left(\frac{x}{4}\right)^2 - \left(\frac{y}{4}\right)^2 \qquad f(x,y) := 1 + J0(6 \cdot x) \cdot e^{-\frac{y}{2}}$$

Choose as initial guess values: $\quad x := 0 \qquad y := 0$

Given $\qquad f(x,y) = g(x,y) \qquad x = 2 \cdot y \qquad \begin{pmatrix} x \\ y \end{pmatrix} := \text{Find}(x,y) \qquad x = 0.405 \qquad y = 0.203$

$$f(x,y) = 0.987 \qquad g(x,y) = 0.987$$

f , g

Figure 3.17:
The completed
worksheet for
Example 3.8

get the small placeholder, ¶ , on the right, bottom, or corner boundary of the graph.

d. Double-click on the graph to get the 3-D Plot Format window. In the QuickPlot Data tab, set the range for Plot 1 to start at 0 end at 2.5, start at 0 end at 5, and # of Grids 100, and for Plot 2 start at 0 end at 2, start at 0 end at 2.5, # of Grids 100 (all of this sets the x and y plotting ranges for the two functions); in the Appearance tab select *Fill Surface, Colormap,* and *No lines*; in the Lighting tab opt for *Enable Lighting*; and in the Advanced tab choose *Colormap Greyscale*. (There are other small details to fix—play around with them!) Click on **OK**.

e. Now the fun begins! Double-click and then hold the mouse button down after doing so—you can now move the mouse around to rotate the graph to any desired orientation.

3. In Fig. 3.17 we show the first of two solutions; to find the second (0.883 and 0.442) vary the initial values for x and/or y.

This completes our discussion of numerically solving an equation or equations for one or more unknowns. To conclude we discuss a few bells and whistles.

Exercises

Note: For some of these problems, you will find Mathcad's built-in Reference Tables very helpful. Click Help . . . Reference Tables.

Note: Be careful with variables and units in these problems! If you do several exercises in the same worksheet, there's a good chance you'll get confusing results (e.g., if you define V as velocity in one problem, you will no longer have V for volts as a built-in unit for a later problem). To avoid this, either use a separate worksheet for each exercise, or make sure you are aware of this and type, for example, **V:=volt** for the second problem.

3.15 Find the loop currents I_1, I_2, and I_3, in the circuit shown, and find the heat generated. Note: Summing voltage drops around each loop, we obtain three equations

$$R_1 I_1 + R_4(I_1 - I_2) - V_1 = 0$$
$$R_2 I_2 + R_4(I_2 - I_1) + V_2 = 0$$
$$R_3 I_3 + V_3 - V_2 = 0$$

The heat generated in a resistor R by current I is RI^2.

3.16 Repeat Exercise 3.15 if V_1 is removed and replaced with a connector (i.e., a zero-resistance connection).

3.17 Repeat Exercise 3.15 if V_2 is removed and replaced with a resistor $R_5 = 10\ \Omega$.

3.18 Find the two points of intersection of a circle radius $r = 5$, centered at $x = 3$, $y = 2$, and a line with slope 2 and intercept 2.

3.19 Find the two points of intersection of a parabola given by $y - 2 = 2(x - 3)^2$, and a line with slope 1 and intercept 2.

3.20 Find the four points (for $x > 0$) of the intersection of the curve $y = 5\sin(\pi x)$, and a line with slope 1 and intercept 1.

3.21 Solve the following equations for w, x, y, and z.

$$\sqrt{w} - x^2 - y + z = -4$$
$$w \cdot y = 12$$
$$y^2 + 4z^2 = 25$$
$$y - z = 5$$
$$z < 0 \quad x > 0 \quad w > y$$

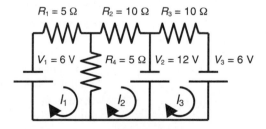

Figure E3.15, E3.16, E3.17

3.22 Find r (m) and θ (°) so that the area of the segment shown in Fig. E3.22 is $A = 11.5$ m², and the total perimeter is $P = 30$ m.

3.23 Find the radius r (in) and the height h (in) of a (reasonable!) circular cylinder if the volume must be $V = 75$ gal and the total surface area (including the top and bottom) must be $A = 18$ ft².

3.24 Find the radius r (in) of the base and the height h (in) of a (reasonable!) right circular cone if the volume must be $V = 0.5$ ft³ and the total surface area (including the base) must be $A = 10$ ft².

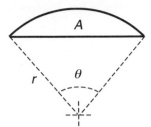

Figure E3.22

3.3

Other Equation-Solving Techniques

Mathcad has a number of other features that can be used with Given. We briefly mention them here, but more detail is available in Mathcad's resources, as shown in Fig. 2.23.

Additional Functions for Use with Given

Mathcad has a range of alternative functions to use in a solve block with Given. Some useful ones are:

1. All the methods we have used are numerical, so a criterion is needed to determine when we have reached sufficient accuracy. This criterion is the value of built-in variable TOL. Its default value is 0.001, but it can be reduced (to increase accuracy at the cost of longer computation time) or increased (to decrease accuracy). To do so, click on *Tools . . . Worksheet Options* and click on the Built-In Variables tab.

2. We have used these methods for numerical solutions, but we will see in Chapter 5 that in some situations they can be used for *symbolic*, or analytic, solutions!

3. Minerr can be used instead of Find, to try and find a solution if Find fails. As the name implies, it tries to gives an approximate solution instead of no solution.

4. Maximize or Minimize can be used instead of Find to optimize a function. In Example 3.4 (Fig. 3.8) we used a differential to minimize a function. Instead we could have used Minimize, as shown in Fig. 3.18. Note that these arguments represent the function you're optimizing (itself *without* arguments) and the independent variable or variables. In general, you may need a constraint equation too.

5. lsolve can be used instead of Find for linear equations (as in Example 3.5). Note also that for linear equations we can use Mathcad's built-in linear algebra capabilities (see Chapter 4).

Choose an initial value for x: $x := 1000 \cdot m$

Given $\underset{\sim}{x} := Minimize(c, x)$ $x = 1207\, m$

The lowest concentration is: $c(x) = 2.86\%$

Press F1 for help.   AUTO   Page 1

Additional Exercises

Note: For some of these problems, you will find Mathcad's built-in Reference Tables very helpful. Click on *Help . . . Reference Tables!*

Note: Be careful with variables and units in these problems! If you do several exercises in the same worksheet, there's a good chance you'll get confusing results (e.g., if you define V as velocity in one problem, you will no longer have V for volts as a built-in unit for a later problem). To avoid this, either use a separate worksheet for each exercise, or make sure you are aware of this and type, for example, **V·=volt** for the second problem!

3.25 The total length of the flight path traveled by a thrown ball is $S = 25$ m, and the distance along the ground is $X = 15$ m. Find the maximum height H of the flight. Note: The length of the flight (parabolic) flight path is given by

$$S = \frac{1}{2}\sqrt{X^2 + 16H^2} + \frac{X^2}{8H}\ln\left(\frac{4H\sqrt{X^2 + 16H^2}}{X}\right)$$

3.26 Using the equation in Exercise 3.25, find the distance X along the ground if the flight path is $S = 201$ ft and the maximum height is $H = 100$ ft.

3.27 A wheel of radius $R = 6$ cm with a point light attached on the edge is rolled along the ground as the motion is videoed. The trajectory of the light is as shown. Find the location y (cm) of the light when the wheel has rolled distances $x = 1$ cm and 10 cm. Note: The trajectory describes a cycloid, for which

Figure E3.27, E3.28

$$x = \cos^{-1}\left(1 - \frac{y}{R}\right) - \sqrt{2Ry - y^2}$$

3.28 Using the equation in Exercise 3.27, find the radius R (cm) of the wheel if the trajectory must pass through point (1 cm, 1 cm).

3.29 The speed V of a space vehicle launched vertically is given by

$$V = U\ln\left(\frac{M}{M - Qt}\right) - gt$$

where $U = 5000$ m/s is the speed of the rocket exhaust (the propulsion), $M = 5 \times 10^5$ kg is the initial vehicle mass, $Q = 3500$ kg/s is the rate of fuel consumption, and t is the time. Find the time at which the spacecraft is moving at 1000 m/s. Note: This formula allows for the fact that as fuel is consumed, the vehicle becomes lighter.

3.30 A vertical rod diameter D = 4 in and length L = 14 ft is to support a load P (lbf). The formula for the largest load is

$$P = \frac{\sigma_{max}\dfrac{\pi D^2}{4}}{1 + \dfrac{4\varepsilon}{D}\sec\left(\dfrac{2L}{D}\sqrt{\dfrac{4P}{\pi E D^2}}\right)}$$

where σ_{max} = 5 × 10⁴ psi is the maximum allowable stress, and $E = 10^6$ psi is the modulus of elasticity, of the rod material, and ε (in) is the eccentricity of the load (i.e., how far from the rod center the load is placed). For example, if ε = 0 in (the load placed dead-center), $P = \sigma_{max}\dfrac{\pi D^2}{4}$. Find the largest load P (lbf) if ε = 0 in, and if ε = 1 in.

3.31 A vertical rod of length L = 14 ft is to support a load P = 105 lbf, placed off-center (eccentricity ε = 1 in). Find the minimum required diameter D (in) that will support the load. Note: Use the equation in Exercise 3.30, with σ_{max} = 5 × 10⁴ psi and $E = 10^6$ psi.

3.32 A supersonic airplane flies at Mach number M = 2.5. The half-angle of the nose of the airplane is θ = 10°. If the atmospheric pressure is p_{atm} = 95 kPa, find the pressure p (kPa and psi) after the oblique shock wave. Note: The pressure after an oblique shock is given by

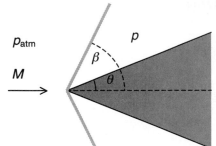

p_{atm}

M

p

β

θ

Figure E3.32

$$p = p_{atm}\left[1 + \frac{2k}{k+1}(M^2\sin^2(\beta) - 1)\right]$$

where k = 1.4 is the ratio of air specific heats and β is the angle of the oblique shock given by

$$\tan(\theta) = 2\cot(\beta)\left[\frac{M^2\sin^2(\beta) - 1}{M^2(k + \cos(2\beta)) + 1}\right]$$

This equation has two solutions for β, corresponding to a strong shock (the larger β) and a weaker one. In supersonic streamlining we wish to only have a weak oblique shock (for a smaller pressure buildup), so use the smaller β.

3.33 The RMS current flowing in the RLC circuit shown is given by

$$I_{RMS}(\omega) = \frac{V_{RMS}}{\sqrt{R^2 + \left(\omega L - \dfrac{1}{\omega C}\right)^2}}$$

where V_{RMS} = 60 V is the RMS voltage, R = 20 Ω is the resistor, L = 500 mH is the inductance, and C = 50 μF is the capacitance. Find the two frequencies ω(Hza) at which the current is 1 A.

3.34 For the circuit shown, with V_{RMS}, R, L, and C as in Exercise 3.33, find the frequency ω(Hza) at which the current I_{RMS} is at its maximum. Note: Find the frequency at which the first derivative of I_{RMS} is zero (with a careful initial guess for ω); then you can compute the maximum current. This frequency is the *tuning* (or *resonant*) *frequency* of the circuit, essentially what you do when you "tune" in a radio station.

3.35 Use the "maximize" function in Mathcad to solve Exercise 3.34.

3.36 A "stiff" spring is one that gets stronger the more it is compressed, unlike a linear spring for which the spring constant is constant. Suppose the force in a stiff spring is defined by $F(x) = kx + \mu x^3$, where x is the amount of compression, $k = 500$ N/m and $\mu = 10,000$ N/m³. Find the compression x (cm) of the spring if a force of 250 N is applied.

3.37 For the spring of Exercise 3.36, if $E = 100$ J is to be stored in the spring, find the compression x (cm) of the spring. Note: The energy stored in a spring is

$$E(x) = \int_0^x F(x)dx.$$

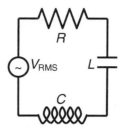

Figure E3.33, E3.34, E3.35

3.38 For the spring of Exercise 3.36, if the *additional* energy required to compress the spring beyond some unknown initial compression to twice that compression is $\Delta E = 200$ J, find the initial compression (cm) of the spring. Note: The energy stored in a spring that is compressed from $x = a$ to $x = b$ is $\Delta E = \int_a^b F(x)dx.$

3.39 Find the values of x between -10 and 10 at which the following function is equal to its own slope: $y(x) = \left(\dfrac{x}{5}\right)^3 - \sin\left(\dfrac{x}{10}\right).$

3.40 Find the values of x between -5 and 5 at which the following function is equal to its own second derivative: $y(x) = e^{-x}\cos(x)$

3.41 Find the positive values of x between 0 and 10 at which the following function is equal to its own integral from 0 to the x value: $y(x) = \left(\dfrac{x}{2}\right)^2 - 3x.$

3.42 A cylindrical storage tank of radius $R = 75$ cm and length $L = 3$ m is partially filled with a liquid to a depth h. If the volume of liquid is (a) 300 gal and (b) 1000 gal, find h (cm). Note: The volume is given by

$$Q(h) = 2L\int_0^h \sqrt{R^2 - (x - R)^2}\, dx.$$

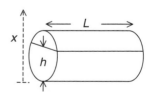

Figure E3.42

3.43 Find the loop currents I_1, I_2, I_3, and I_4, in the circuit shown at the top of page 58, and find the heat generated. Note: Summing voltage drops around each loop, we obtain three equations

$R_1 I_1 + R_5(I_1 - I_2) - V_1 = 0$

$R_2 I_2 + R_5(I_2 - I_1) + V_2 = 0$

$R_3 I_3 + R_6(I_3 - I_4) - V_2 = 0$

$R_4 I_4 + R_6(I_4 - I_3) + V_3 = 0$

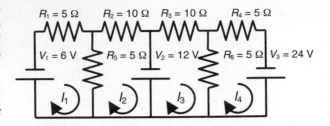

The heat generated in a resistor R by current I is RI^2.

3.44 Repeat Exercise 3.43 if V_1 is removed and replaced with a connector (i.e., a zero-resistance connection).

3.45 Repeat Exercise 3.43 if V_2 is removed and replaced with a resistor $R_7 = 10\ \Omega$.

3.46 Find all the points between $x = 0$ and $x = 50$ at which $y(x) = \left(1 - \dfrac{x}{10}\right)^3$ and $y(x) = x\cos\left(\dfrac{x}{20}\right)^2$ (not $(\cos)^2$) intersect.

3.47 Solve the following equations for x, y, and z.

$$x \cdot z = -6$$
$$x + \sqrt{y} = 5$$
$$x^2 + y = 13$$

3.48 What is the flow rate Q (gal/min) through $L = 2000$ ft of water pipe, with pipe diameter $D = 1$ in, generated by a pump that can produce a pressure of $\Delta p = 25$ psi? Note: The flow rate is given by $Q = \dfrac{\pi D^2}{4} V$, so we need to find the velocity V of water in the pipe. It turns out that to find V we need to solve simultaneously for it and also the Reynolds number Re, and friction factor f. The set of equations to solve for V, Re, and f are

$$\Delta p = f\frac{L}{D}\frac{V^2}{2}$$

$$Re = \frac{\rho VD}{\mu}$$

$$\frac{1}{\sqrt{f}} = -2\log\left(\frac{e/D}{3.7} + \frac{2.51}{Re\sqrt{f}}\right)$$

where ρ and μ are the density and viscosity of water and $e = 0.01$ in is the roughness of the pipe surface. Use $V = 1$ ft/s as your initial guess, and compute from it an initial guess value for Re; for f use an initial guess less than 1.

Vectors, Matrices, and More

How to Create Vectors and Matrices in Mathcad

We've already introduced how to create vectors and matrices in Mathcad in previous chapters. Let's review all the details of how to do this. There are three basic ways to create vectors (which are simply matrices with a single column) or matrices:

1. Use the Insert Matrix window shown in Fig. 4.1. To access this there are three methods: using *Insert ... Matrix*; using the accelerator keys **Ctrl + M**; or using the Matrix or Vector icon (▦) on the Vector and Matrix Toolbar (available from the Math Toolbar—if this is not visible use *View ... Toolbars ... Math*). Once you have this window, simply select the desired row and column size of the matrix (e.g., a vector will have one column). Then you can type in the cell values. To go from cell to cell, either use the mouse or the Tab key.

2. Use Mathcad's array subscript notation. This is tricky until you get used to it because there are a couple of errors that beginners frequently make. We'll discuss this approach in detail in the next subsection.

3. Use one of several methods for importing data (e.g., from Excel or Word), including simply copying and pasting. We will discuss this method in detail in Chapter 7.

For now, you can practice the first method by creating the vectors and matrices shown in Fig. 4.2.

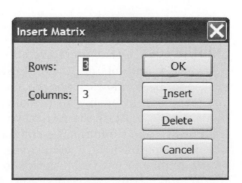

Figure 4.1: The Insert Matrix window

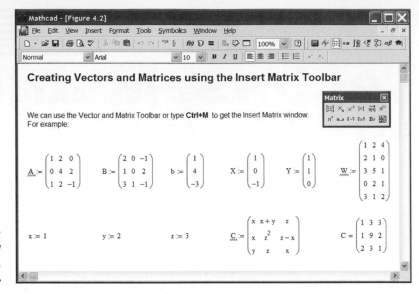

*Figure 4.2:
Creating vectors and
matrices using the
Insert Matrix window*

Mathcad's Two Subscript Notations

The following discussion applies whether you are defining (by typing **:**) or evaluating (by typing **=**) a vector or matrix variable.

It's very important to be clear about what kind of subscript you wish to create. There are two different kinds in Mathcad, one of which we've already seen in previous chapters. Let's review in detail what these two are (and for extra practice you might want to replicate what follows in a practice worksheet):

1. *Literal subscripts*. As we saw in previous chapters, these are subscripts you create for your own convenience in naming constants or variables. To create these you must type a period (**.**). For example, in Mathcad you could type **x.2** and **y.max**. (Press **Enter** after each one.)

2. *Array subscripts*. These are subscripts used for creating or evaluating vector or matrix elements. They are generated by typing a left square bracket, **[** (or by using the $\times_n$ icon on the Vector and Matrix Toolbar). For example, in Mathcad you could type **x[2**, **y[max**, and **z[i,j**. (Press **Enter** after each one.)

This somewhat quirky use of the period and square bracket is a convention unique to Mathcad. It is, at first glance, a bit odd, until you realize that there are only a limited number of keys on a keyboard, and lots of accelerator-key combinations are already spoken for by Windows. If you wish to use these subscript notations, you just have to memorize these two particular keystrokes. The last example, **z[i,j**, is the format you would use for specifying the cell in row i and column j of a matrix z.

Unless you have very young eyes, it's quite difficult (depending on the fonts you use in Mathcad) to visually distinguish between literal and array subscripts; Fig. 4.3 shows our results. *Even though they look alike, they have completely different meanings*: the literal subscript is simply a way to label a variable (e.g., it would be convenient to call the maximum height reached by a projectile y_{max}); the array subscript indicates which cell of a vector or matrix is being referred to (e.g., x_2 refers to the *third* cell in a vector x).

Literal subscripts look like: x_2 y_{max} $z_{i,j}$

Array subscripts look like: x_2 y_{max}

*Figure 4.3:
Literal and array
subscripts*

Let's repeat that comment in parentheses: the array subscript in x_2 refers to the *third* cell in a vector x. This is because by default Mathcad counts elements in a vector or matrix starting with zero. Mathcad has a built-in variable, ORIGIN, which is by default zero, for setting how elements in a vector or matrix are counted. If you wish to change its value, you can do either of the following:

1. Type at any point in a worksheet a new value for ORIGIN (e.g., **ORIGIN:1**). From that point on, the first element's subscript in a vector will be $_1$, not $_0$, and in a matrix will be $_{1,1}$, not $_{0,0}$.
2. Change the ORIGIN value for the entire worksheet by using *Tools . . . Worksheet Options* and clicking on the Built-In Variables tab.

It's also important to note that if we first define an element by typing, say, **x[2:π**, to generate $x_2 := \pi$, we have not only defined x_2 but also told Mathcad that we have a vector x whose *last* element is $x_2 = \pi$. If we haven't previously defined the other elements of x, Mathcad creates them and assigns their values to zero!

This all sounds a bit complicated, but some examples, as shown in Fig. 4.4 (which you should try to reproduce), will clarify things. In all expressions we used the array subscript. The default value of ORIGIN is zero, so when we define $x_2 = \pi$, we also automatically create $x_0 = 0$ and $x_1 = 0$. A similar thing happens when we create $z_{3,2} = 5$. We then use method 1 above to change ORIGIN to 1. When we ask Mathcad for x, it knows it's a three-dimensional vector, but it reassigns the subscripts, so that the lowest is $_1$, x_0 doesn't exist any more, and x_2 is now zero! A similar thing happens with z. As an exercise, see if you can anticipate what will happen if you now change ORIGIN to 1 using method 2 above.

This demonstrates the effect of the ORIGIN value. You should be always aware of the value of ORIGIN in each of your worksheets; you should not change it in the middle of any worksheet (i.e., use method 1 above) unless you have some good reason for doing so.

This is a good place to review Mathcad's notation for creating and using range variables.

Defining Ranges in Mathcad

Mathcad uses the notion of ranges for several purposes: to define arrays, to plot functions, to do iterative calculations, and to evaluate certain control sums and products. A range is nothing more than a

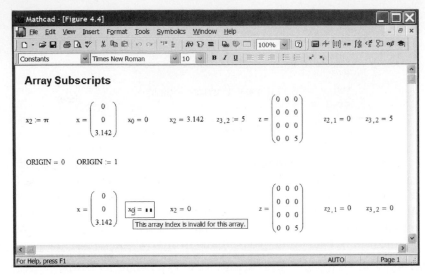

sequence of numbers that you can define; some examples are shown in Fig. 4.5, which you can try to reproduce as we discuss defining ranges. Note that the results you obtain may display differently than those in Fig. 4.5; Mathcad will display arrays as a matrix or table (spreadsheet) depending on the array size and your settings under *Format . . . Result*.

To define a range in Mathcad you can:

1. Type **a;b** where a and b are the beginning and end of the range (they must be real, but can be integers, decimals, fractions, or previously defined variables). Mathcad replaces the semicolon (;) with two dots (..); *you cannot type the two dots*! (Instead of typing **;** you can get them by using the ▣ icon on the Vector and Matrix Toolbar) In Fig. 4.5 the ranges i and j are

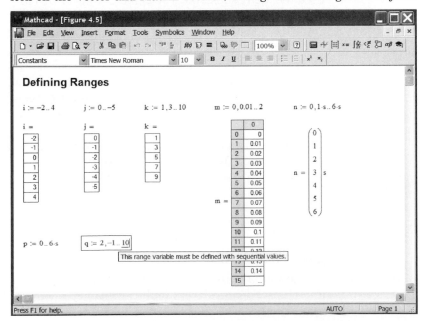

**Figure 4.5:
Some examples
of ranges**

examples of this. Note that we can generate an increasing *or* decreasing range. Regardless of the values of a and b, Mathcad fills in the range between the two *using a step size of unity* (e.g., for an increasing range, a, $a + 1$, $a + 2$, until we reach b). The last value will be less than or equal to b, if a and b are decimals.

2. Type **a,b;c** where a is the beginning value, b is the second value in the series, and c is the end of the range (with the same constraints as in step 1). Mathcad again replaces the semicolon (;) with two dots (..). In Fig. 4.5 the ranges k and m are examples of this. Mathcad computes the required step size from the supplied values. *The most common error in using this approach is to assume that the middle value, b, represents the step size rather than the second value* (in range k the step size is 2, not 3)! Note that if you wish to define a range with units (as in range n in Fig. 4.5), you *must* use this three-number format, and you obviously must use consistent dimensions with the three values.

Figure 4.5 also shows a couple of common errors. For range p we try to define a range with units without specifying the middle term; range q is invalid because you cannot start at 2, have a second term of -1 (creating a step size of -3), and end up at 10.

We can mention a couple of subtleties in defining ranges with units. As in range n, we can use zero without units as a value in a range that has units, although it's good practice to define it with units; and the displayed results will use the underlying fundamental units. One of the most important points is that *all equations that follow and use the range variable are immediately evaluated for all values of the range.*

As previously stated, ranges can be used for several purposes. Figure 4.6 shows some of these—try reproducing it. Some specific examples are:

1. Defining arrays (vectors and matrices). (See V and M in Fig. 4.6).

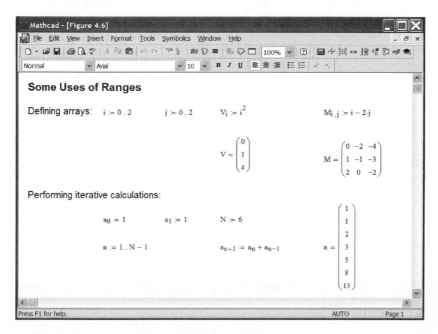

Figure 4.6: Examples of using ranges

2. Controlling the appearance of graphs. (See Appendix: Graphing.)
3. Doing iterative calculations. After defining a range (which in this case must be integers), *all subsequent equations that use the range variable as a subscript will immediately be computed for all values of the range.* (See Fig. 4.6, where we use iteration for computing a few values of the Fibonacci numbers, defined as 1, 1, then an infinite sequence in which each term equals the sum of the previous two terms.)

We will now do some examples of using ranges.

Example 4.1: Iterating a Range with the Newton-Raphson Method
The well-known Newton-Raphson method can be used to find the roots of a function $f(x)$. (Yes, it's much more convenient to use Mathcad's root or Given ... Find to find roots!) In this method, we start with an initial guess, x_0, as shown in Fig. 4.7. If we construct the tangent to the curve at this point, it will intersect the x-axis at a point, x_1, that will (usually) be closer to the root we're looking for. We can repeat this method to find x_2, x_3, and so on, until we are satisfied that the change in x is negligible. The equation shown in Fig. 4.7 is the mathematical expression of this method.
Find the roots of

$$f(x) = 2x^3 + 3x^2 - 18x - 27$$

The solution is shown in Fig. 4.8. You should try to reproduce it and to find the other two roots by changing the initial guess value, x_0.

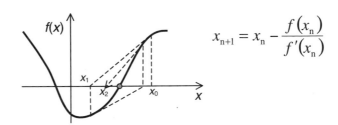

Figure 4.7:
The Newton-Raphson
method concept

$$x_{n+1} = x_n - \frac{f(x_n)}{f'(x_n)}$$

Example 4.2: The Taylor Series: Summing over a Range Compare the function $e^{-\frac{x}{2}}$ to its Taylor series representation

$$f(x) = \sum_m \frac{\left(-\frac{x}{2}\right)^2}{m!}$$

(with $m = 0, 1, 2, \ldots, N$) by evaluating both at $x = 5$. Start with $N = 1$ and then change to 2, 3, 4, and so on, to find how many terms (N) are needed to make the error between the two less than 1% at $x = 5$.

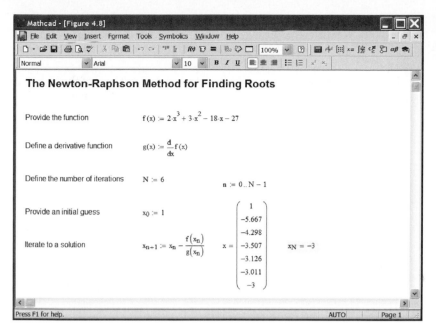

Figure 4.8:
The Newton-Raphson
method for
Example 4.1

The worksheet is shown in Fig. 4.9. After reproducing the worksheet, change *N* until the value of *Error* is less than 1% (notice how the graph changes). Some suggestions for reproducing the worksheet are:

1. The summation icon ($\sum_{n}$) can be obtained from the Calculus Toolbar or by typing **Shift + 4**.

2. The factorial can be obtained from the Calculator Toolbar or by simply typing **!**.

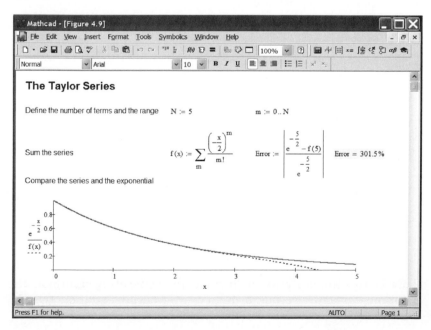

Figure 4.9:
The Taylor series

3. The error is defined as shown; the absolute value (| |) can be obtained from the ⋈ icon on the Calculator Toolbar (*Not* the one on the Matrix Toolbar—that's for computing the determinant of a matrix!), or by typing **Shift + **; the "units" of % were inserted.

4. Generating graphs is discussed in detail in Appendix: Graphing, but to reproduce the one shown:

> **a.** Click on *Insert . . . Graph . . . X-Y Plot* (or use the ⬓ icon on the Graph Toolbar) or type **@**.
>
> **b.** Click on each of the three black placeholders on the horizontal axis and type (starting with the one on the left), **0**, **x**, and **5**.
>
> **c.** Click on middle black placeholder of the vertical axis and type **e^-x/2 Spacebar Spacebar , f(x)**.
>
> **d.** Press **Enter** to exit the graph.
>
> **e.** You can double-click on the graph to make the axis style crossed and to format the curves.
>
> **f.** To move the graph click once on it and move the mouse to the edge, where you'll get a small black hand, ✋ ; use this to move the graph.
>
> **g.** You can resize the graph by clicking once on it and dragging on the small placeholder, ◂ , on the right, bottom, or corner boundary of the graph.

Example 4.3: Euler Method for a First-Order ODE Use the Euler method to solve the following differential equation:

$$\frac{dy}{dx} = 2x - y \quad y(0) = 4$$

over the range $x = 0$ to $x = 5$, using $N = 25$ steps. Compare to the exact solution $y_{exact}(x) = 6e^{-x} + 2x - 2$ by evaluating both at $x = 5$. Then explore what happens as you increase N.

The Euler method uses the following approximation for the slope of a curve at some point (x_n, y_n) (see Fig. 4.10):

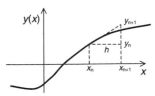

**Figure 4.10:
The Euler Method**

$$\frac{dy}{dx} \approx \frac{y_{n+1} - y_n}{x_{n+1} - x_n}$$

If we choose a *step size* $h = x_{n+1} - x_n$, then the above equation can be combined with a differential equation of form

$$\frac{dy}{dx} = f(x,y)$$

to give us

$$y_{n+1} \approx y_n + hf(x_n, y_n)$$

This is the Euler iteration formula for generating approximate solutions to a first-order ordinary differential equation (ODE). Note that this method is very simple, but is also not very accurate.

In Chapter 6 we will revisit this method and use built-in Mathcad functions to much more accurately solve ODEs. You can go ahead and reproduce the worksheet shown in Fig. 4.11. Note the following suggestions:

1. The differential equation is just for us to look at—Mathcad does not use it or the initial condition.

2. We set up the math so that we always compute from $x = 0$ to $x = L$ (= 5); if we change the number of steps N, the step size h changes, but not L.

3. All subscripts except $_{exact}$ are array subscripts.

4. The range n goes up to $N - 1$ so that x_N and y_N are the last terms computed.

5. Generating graphs is discussed in detail in Appendix: Graphing, but to reproduce the one shown:

   a. Click on *Insert . . . Graph . . . X-Y Plot* (or use the 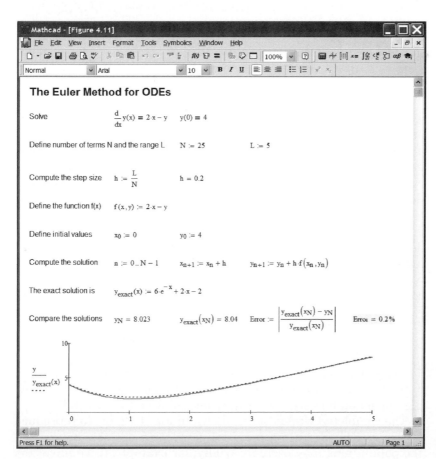 icon on the Graph Toolbar) or type **@**.

   b. Click on each of the three black placeholders on the horizontal axis and type (starting with the one on the left), **0**, **x**, and **5**.

   c. Click on middle black placeholder of the vertical axis and type **y, y.exact(x)**.

   d. Press **Enter** to exit the graph.

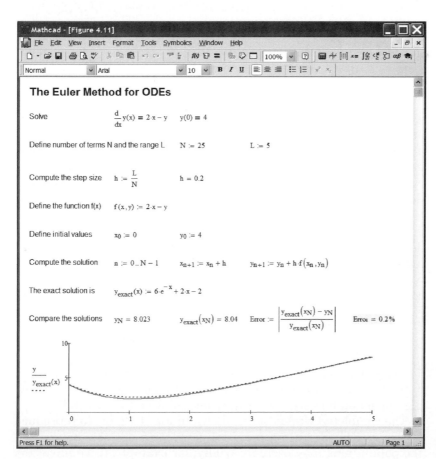

Figure 4.11:
The Euler method
for ODEs

e. You can double-click on the graph to make the axis style crossed and to format the curves.

f. To move the graph click once on it and move the mouse to the edge, where you'll get a small black hand, 🖐 ; use this to move the graph.

g. You can resize the graph by clicking once on it and dragging on the small placeholder, ◂, on the right, bottom, or corner boundary of the graph.

The next example illustrates a very important concept when computing several equations that all use the same range.

Example 4.4: Euler Method for Simultaneous (Coupled) First-Order ODEs An island is to be seeded with an initial population of deer and wolves, in an experiment in population growth. The deer subsist on grass (essentially an unlimited supply), and the wolves subsist on deer. Predict the populations of deer and wolves over 4 years. The natural growth rates of deer and wolves can be approximated as 1 and 0.1, and the natural death rates of deer and wolves can be approximated as 0.05 and 10, respectively, in the Lotka-Volterra equations.

This is a classic problem in population studies, and is modeled using the Lotka-Volterra equations

$$\frac{dD}{dt} = g_\mathrm{D}D - d_\mathrm{D}D \cdot W$$

$$\frac{dW}{dt} = g_\mathrm{W}D \cdot W - d_\mathrm{W}W$$

where D and W are the number of deer and wolves, respectively, at any time t, g_D, and g_W are the natural growth rates of the deer and wolves, and d_D and d_W are the natural death rates of the deer and wolves. The idea behind these equations is that the growth rate of deer depends directly on the number of deer (i.e., more deer will have more offspring), and their death rate depends on the probability of an encounter between a deer and a wolf (hence deaths depend on the product $D \cdot W$). Similarly, the wolf growth rate depends on this probability, and the death rate depends on natural factors but is proportional to how many wolves are present (i.e., we can assume a fixed percentage of the wolves die per year). The corresponding Euler equations are

$$D_{\mathrm{n}+1} = D_\mathrm{n} + \Delta t(g_\mathrm{D}D_\mathrm{n} - d_\mathrm{D}D_\mathrm{n}W_\mathrm{n})$$
$$W_{\mathrm{n}+1} = W_\mathrm{n} + \Delta t(g_\mathrm{W}D_\mathrm{n}W_\mathrm{n} - d_\mathrm{W}W_\mathrm{n})$$

These equations are coupled—that is, we can't solve one, then the other, but rather must solve them simultaneously. This will have a big impact on how we must set up our solution in Mathcad!

The completed worksheet is shown in Fig. 4.12; try to reproduce it. We see that for a given initial population of deer and wolves the populations may oscillate wildly; for example, if we have "too

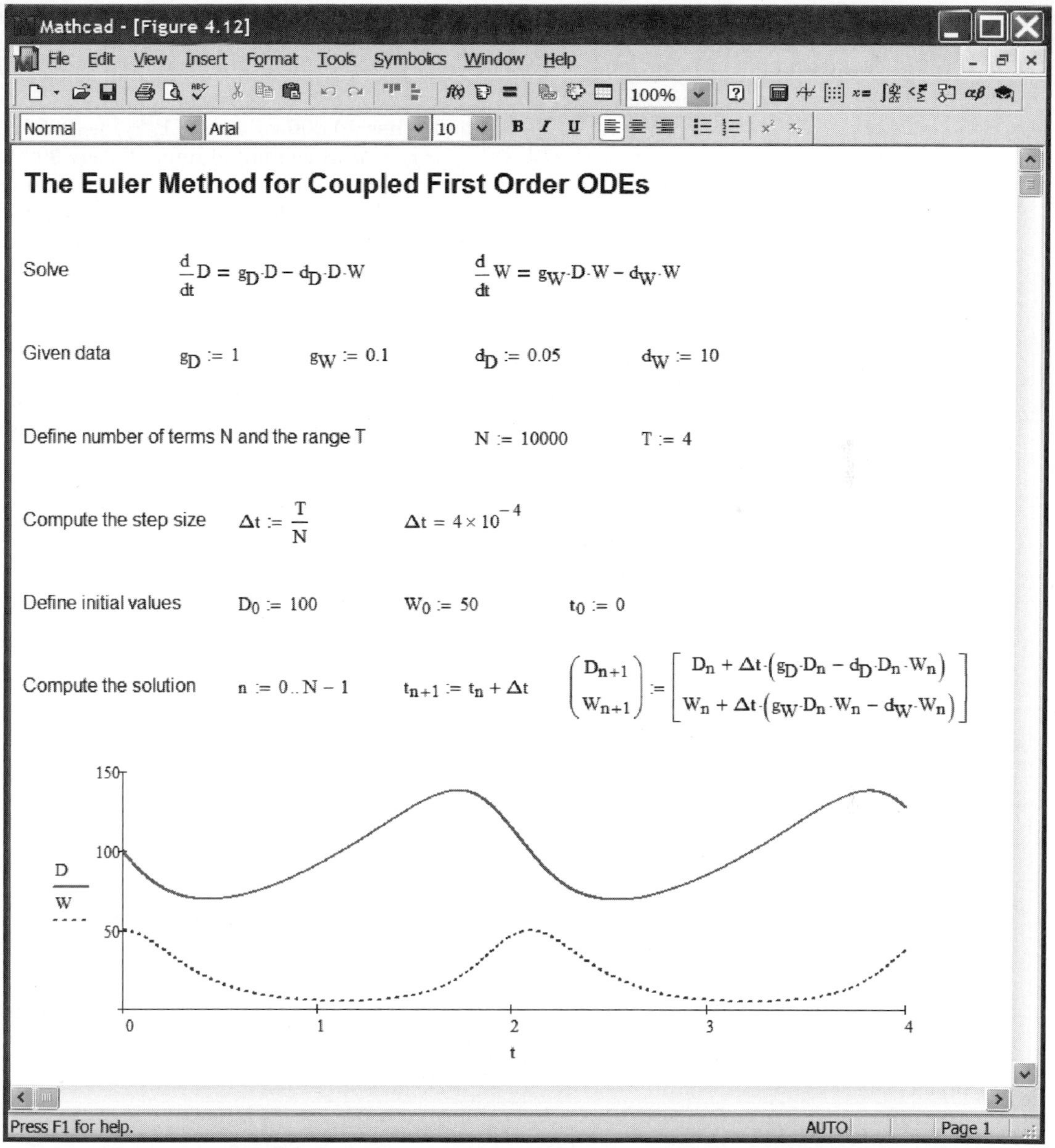
Mathcad - [Figure 4.12]
File Edit View Insert Format Tools Symbolics Window Help
Normal
Arial
10
The Euler Method for Coupled First Order ODEs
Solve
$\frac{d}{dt}D = g_D \cdot D - d_D \cdot D \cdot W$
$\frac{d}{dt}W = g_W \cdot D \cdot W - d_W \cdot W$
Given data
$g_D := 1$
$g_W := 0.1$
$d_D := 0.05$
$d_W := 10$
Define number of terms N and the range T
$N := 10000$
$T := 4$
Compute the step size
$\Delta t := \frac{T}{N}$
$\Delta t = 4 \times 10^{-4}$
Define initial values
$D_0 := 100$
$W_0 := 50$
$t_0 := 0$
Compute the solution
$n := 0..N-1$
$t_{n+1} := t_n + \Delta t$
$\begin{pmatrix} D_{n+1} \\ W_{n+1} \end{pmatrix} := \begin{bmatrix} D_n + \Delta t \cdot (g_D \cdot D_n - d_D \cdot D_n \cdot W_n) \\ W_n + \Delta t \cdot (g_W \cdot D_n \cdot W_n - d_W \cdot W_n) \end{bmatrix}$
150
100
50
D
W
0
1
2
3
4
t
Press F1 for help.
AUTO
Page 1

many" wolves, the deer population will plunge, followed by a plunge in the wolf population (it is feast or famine for the wolves), leading to a resurgence in the deer population, and so on. It turns out that if we choose the two initial populations wisely, both populations will be very stable—try different values of D_0 and W_0 to try to find this. (Hint: Set the left-hand sides of the original Lotka-Volterra equations to zero and solve for D and W.)

Figure 4.12:* *The Euler method for coupled first-order ODEs

We make the following points:

1. The differential equations and initial conditions are just for us to look at—Mathcad does not use them.

2. All subscripts except $_D$ and $_W$ are array subscripts.

3. The rest of the worksheet is fairly straightforward, except for the following very important point: *Whenever you define a range in Mathcad, any and all subsequent equations that contain the range variable are immediately computed for all instances of that variable*. In our example, the range variable is n, and it has 10,000 values (0, 1, 2, . . . , 9999), so the equation for the times, $t_{n+1} = t_n + \Delta t$, is computed immediately 9999 times, to compute all the time steps. More importantly, it means we must put the simultaneous equations in a matrix form, as shown! If we wrote the two equations as two separate equations, the computation would fail! The reason for this is as follows. Consider what would happen if we had written the equation $D_{n+1} = D_n + \Delta t(g_D D_n - d_D D_n W_n)$ without using the matrix, as shown in Fig. 4.13.

For $n = 0$ we get $D_1 = D_0 + \Delta t(g_D D_0\ 2\ d_D D_0 W_0)$, which does compute; for $n = 1$ we get $D_2 = D_1 + \Delta t(g_D D_1 - d_D D_1 W_1)$, which does *not* compute because we don't yet have W_1; hence all subsequent values of D fail to compute! If we tried computing the W equation first, we would get a similar problem. This difficulty of trying to compute from two or more simultaneous coupled equations over a range is one of the most common mistakes when doing iterative calculations!

The correct approach is shown in Fig. 4.14. For each value of n we compute the matrix equation once; D_0 and W_0 lead to D_1 and W_1, D_1 and W_1 lead to D_2 and W_2, and so on.

4. Generating graphs is discussed in detail in Appendix: Graphing, but to reproduce the one shown:

a. Click on *Insert . . . Graph . . . X-Y Plot* (or use the ⤢ icon on the Graph Toolbar) or type **@**.

b. Click on the middle black placeholder on the horizontal axis and type **t**.

c. Click on middle black placeholder of the vertical axis and type **D, W**.

d. Press **Enter** to exit the graph.

e. You can double-click on the graph to make the axis style crossed and to format the curves.

f. To move the graph click once on it and move the mouse to the edge, where you'll get a small black hand, ✋ ; use this to move the graph.

We will have much more to say about differential equations in Chapter 6, including discussing several functions built in to Mathcad that are very useful for solving such equations.

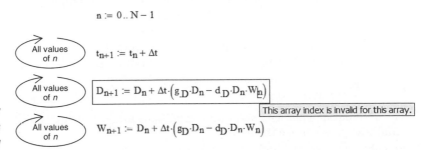

**Figure 4.13:
*Iterative calculations
gone wrong***

$$n := 0 .. N - 1$$

All values of n

$$t_{n+1} := t_n + \Delta t$$

All values of n

$$\begin{pmatrix} D_{n+1} \\ W_{n+1} \end{pmatrix} := \begin{bmatrix} D_n + \Delta t \cdot \left(g_D \cdot D_n - d_D \cdot D_n \cdot W_n \right) \\ W_n + \Delta t \cdot \left(g_W \cdot D_n \cdot W_n - d_W \cdot W_n \right) \end{bmatrix}$$

*Figure 4.14:
Iterative calculations
done correctly*

Exercises

4.1 Define and display a range x that goes from -2π to 20π, in steps of 2π.

4.2 Define and display a range x that goes from 3 in to 2 ft, in steps of 3 in.

4.3 Define and display (in units of ft²) a range A that goes from 0 to 1.5 m², in steps of 0.25 m².

4.4 Define and display an x range that has values 27, 22, 17, 12, 7, and 2.

4.5 Define $x_{start} = -3$, $x_{end} = 6$, and $\delta = 1.5$. Then define (in terms of x_{start}, x_{end}, and δ) and display a range x that starts at x_{start} and ends at x_{end}, with step size δ. Then change δ to 0.75 and display the new x values.

4.6 Define $N = 20$; then define ranges $i = 0 - 5N$ and $j = 0 - 5N$. Then define the array M as $M_{i,j} = \sin\left(\dfrac{i}{N}\right)\cos\left(\dfrac{j}{N}\right)e^{-\frac{i}{5N}}$. Find the maximum and minimum value of M (just type **max(M)=** and **min(M)=**). Plot M on a surface plot and work to make it pretty (e.g., color). Refer to Appendix: Graphing for help on this. Explain the effect that changing the value of N has on the contents of the worksheet.

4.7 Evaluate $\sum\limits_{k=0}^{n}\dfrac{1}{k} - \ln(N)$, for $N = 50$, two ways: (a) using the icon on the Calculus Toolbar and (b) defining a range $i = 1 - N$, defining $x_1 = 0$, performing the iterative calculation $x_{i+1} = x_i + \dfrac{1}{i}$, and then evaluating x_{N+1} $- \ln(N)$. In each case, we are summing the series $1 + \dfrac{1}{2} + \dfrac{1}{3} + \dfrac{1}{4} + \dfrac{1}{5} + \ldots$ before subtracting $\ln(N)$. Explain why we use x_{N+1} and not x_N in case (b). Examine the effect on the two evaluations if we change N to 100, 1000, and 10,000. Look up Euler's constant online!

4.8 Evaluate $\sum\limits_{k=1}^{N}\dfrac{1}{k!} = \dfrac{1}{0!} + \dfrac{1}{1!} + \dfrac{1}{2!} + \dfrac{1}{3!} + \ldots + \dfrac{1}{N!}$, for $N = 5$, two ways:

(a) using the icon on the Calculus Toolbar (and you can just type ! for the factorial function) and (b) defining a range $i = 0 - N$, defining $x_0 = 0$, performing the iterative calculation $x_{i+1} = x_i + \dfrac{1}{i!}$, and then evaluating x_{N+1}. Try increasing N to 10, 20, and so on. What is the meaning of the result we obtain? (You should recognize it.) Explain why we use x_{N+1} and not x_N in case (b).

4.9 Evaluate $4\sum_{n=0}^{n}\dfrac{(-1)^n}{2n+1} = 4\left(\dfrac{1}{1} - \dfrac{1}{3} + \dfrac{1}{5} - \dfrac{1}{7} + \dfrac{1}{9} - \ldots\right)$, for $N = 100$,

two ways: (a) using the ∑ icon on the Calculus Toolbar and (b) defining a range $i = 0 - N$, defining $x_0 = 0$, performing the iterative calculation $x_{i+1} = x_i + \dfrac{(1)^i}{2i+1}$, and then evaluating $4x_{N+1}$. Try increasing N to 500, 1000, 10,000, and so on. What is the meaning of the result we obtain? You should recognize it. Explain why we use x_{N+1} and not x_N in case (b).

4.10 Define $y(x) = \sum_{m=0}^{N}\dfrac{(-1)^m}{k!m!}\left(\dfrac{x}{2}\right)^{2m}$ for $N = 5$ (use the ∑ icon on the Calculus Toolbar and you can just type ! for the factorial function). Plot it and $J_0(x)$ (the Bessel function of the first kind, order zero) on the same X-Y graph. For the graph, click *Insert ... Graph ... X-Y Plot* (or use the ⊾ icon on the Graph Toolbar) or type @; on each of the three black placeholders on the horizontal axis, type (starting with the one on the left) **0, x**, and **10**; on each of the three black placeholders on the vertical axis, type (starting with the one on the bottom) **−1**, and **y(x),J0(x)**, and **1**. To spruce up the graph, refer to Appendix:Graphing. Explain what happens to the graph as you increase or decrease the value of N. The Bessel function appears in many engineering problems, especially problems that deal with cylindrical coordinates (e.g., the graph could represent the instantaneous surface of a vibrating drumskin; x in this case represents the radial position r).

4.11 Use the Euler method to solve $\dfrac{dy}{dx} = xy^2$, $y(0) = 1$ from $x = 0$ to $x = 1.25$. Written as an Euler equation, this is $y_{n+1} = y_n + hf(x, y)$, with $f(x, y) = xy^2$, and h the step size. Choose $N = 5000$ steps. Compare your solution to the exact one, $y_{Exact}(x) = \dfrac{2}{2 - x^2}$, by plotting both on an X-Y graph. For the graph click *Insert ... Graph ... X-Y Plot* (or use the ⊾ icon on the Graph Toolbar) or type @; on the middle black placeholder on the horizontal axis type x; on the middle black placeholder on the vertical axis type **y,y.Exact(x)**. To spruce up the graph, refer to Appendix:Graphing.

4.12 A spherical dust particle (density $\rho_p = 7000$ kg/m³, radius $r = 0.05$ mm) falls from rest in water (density $\rho = 1000$ kg/m³, viscosity $\mu = 10^{-3}$ N·s/m²). Using the Euler method (with units), plot the particle velocity versus time as it falls under gravity. What is the terminal (maximum) velocity? Compare your result to the theoretical maximum of $V_{max} = a/b$. The equation of motion for this is $\dfrac{dV}{dt} = a - bV$, where $a = g\dfrac{\rho_p - \rho}{\rho_p}$ and $b = \dfrac{6\pi\mu r}{M}$, with M being the particle mass. Written as an Euler equation, this is $V_{n+1} = V_n + \Delta t(a - bV_n)$, with time steps given by $t_{n+1} = t_n + \Delta t$. Choose a time range of $T = 0.025$ s, with $N = 5000$ steps.

4.2

Vector Math

Mathcad can perform the math required when working with three-dimensional vectors, with or without units. Some of this is shown in Fig. 4.15. You can try to reproduce it, with the following suggestions:

1. We created a special format for vectors (you do not need to do this step to do vector math, but it does make the screen look nice). To create this format click on *Format . . . Equation* to get an Equation Format window showing current math region styles (*Variables, Constants*, and a group of user styles); you can select one of the user styles and rename it as *Vector*, modified to be bold. When you now type a variable, you can apply this math style to it using the Formatting Toolbar. Note that Mathcad now treats *a* and *a* as different variables!

2. We changed the value of ORIGIN to unity using *Tools . . . Worksheet Options* and clicking on the Built-In Variables tab so that the first element in a vector has the subscript $_1$ not $_0$. The three subscripts in Fig. 4.15 are array subscripts.

3. The dot product and the cross product are obtained from the Vector and Matrix Toolbar or by typing the accelerator keys (e.g., simply * for the dot product and **Ctrl + 8** for the cross product); the magnitude symbol | | must be obtained from the |x| icon on the Calculator Toolbar (*not* the one on the Vector and Matrix Toolbar, which as mentioned previously, is for computing the determinant of a matrix) or by typing **Shift + **.

4. There are three non-standard operations shown in Fig. 4.15 that you can do in Mathcad: we are allowed to add a scalar to a vector, as shown in the expression $a + b$ (Mathcad adds the scalar to all elements of the matrix); we can sum the elements of a matrix; and we can *vectorize* an operation (by using the icon on the Vector and Matrix Toolbar or by typing **Ctrl + -**), which causes the operation to be done on each element of the vectors, as shown.

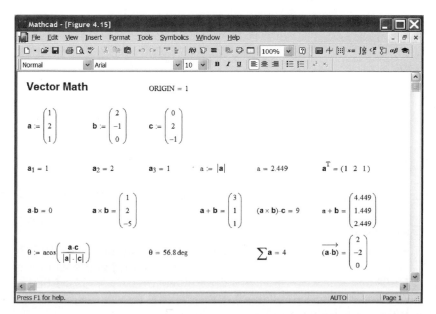

Figure 4.15: Some vector math

Example 4.5: Vector Math with Units: Work Done by, and Torque of, a Force A three-dimensional force is given by $F = i10 + j5 + k15$ (N). It is initially applied at $x_1 = i2 + j3 + k1$ (m). Find the torque it produces at the origin. Now the force moves to $x_2 = \mathbf{i}5 + \mathbf{j}3 + \mathbf{k}5$ (m). Find the torque it now produces at the origin. Find the total distance moved, $\Delta x = x_2 - x_1$, and the total work done.

Recall that torque produced by a force $\boldsymbol{F}$ at location $\boldsymbol{x}$ is $\boldsymbol{T} = \boldsymbol{x} \times \boldsymbol{F}$, and the work when the force moves along path $\Delta \boldsymbol{x}$ is $W = \boldsymbol{F} \cdot \Delta \boldsymbol{x}$. The solution is shown in Fig. 4.16, which you should reproduce. Note the following:

1. Although we could have directly defined $\boldsymbol{F}$, $\boldsymbol{x}_1$, and $\boldsymbol{x}_2$, we decided to define unit vectors $\mathbf{i}, \mathbf{j}$, and $\mathbf{k}$, and use these to build the vectors.

2. The subscripts for $\boldsymbol{x}_1$ and $\boldsymbol{x}_2$ are literal subscripts.

3. We again created a special bold format for vectors (because it looks nice).

4. For help on dot and cross products and the absolute value symbol, see point 3 in the previous list.

5. We inserted units of $N \cdot m$ in the torques to replace J, Mathcad's choice.

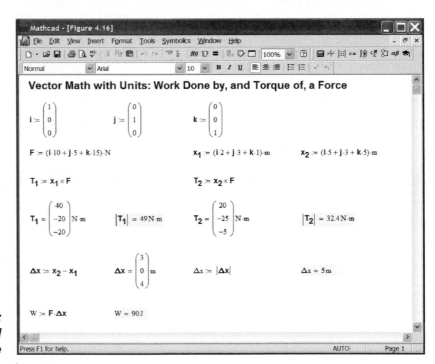

*Figure 4.16:
Work done by, and
torque of, a force*

Exercises

For these exercises, create and use a Vector font, and set ORIGIN to 1 as discussed above, and define unit vectors $\mathbf{i}, \mathbf{j}$ and $\mathbf{k}$ as in Fig. 4.16. The vectors $\mathbf{a} = \mathbf{i}3 + \mathbf{j}5 + \mathbf{k}2$, $\mathbf{b} = \mathbf{i}4 + \mathbf{j}2 + \mathbf{k}2$, $\mathbf{c} = \mathbf{j}2 + \mathbf{k}3$, and the force and position vectors $\mathbf{F} = (\mathbf{i}1 + \mathbf{j}2 + \mathbf{k}1)$ N, and $\mathbf{r} = (\mathbf{i}4 + \mathbf{j}2 + \mathbf{k}2)$ m are also used in some problems.

4.13 Find the angles (°) between $\mathbf{a}$ and $\mathbf{b}$, and between $\mathbf{a}$ and $\mathbf{c}$.

4.14 Find the length of vector $\mathbf{d} = \mathbf{a} + \mathbf{b} + \mathbf{c}$, and the angles (°) it makes with $\mathbf{a}, \mathbf{b}$, and $\mathbf{c}$.

4.15 Find the volume defined by vectors **a**, **b**, and **c** by computing the triple products **a**·(**b** × **c**) and (**a** × **b**)·**c**.

4.16 Find a unit vector perpendicular to **a** and **b**.

4.17 Find the magnitude of force **F** (in lbf and in N) and the magnitude of the torque **T** (N·m and lbf·ft) it produces at a radius arm of **r**.

4.18 Find the work done (J and lbf·ft) if **F** is moved along vector distance **r**.

4.19 Find the angles (°) that the force **F** and the position **r** make with the x-, y-, and z-axes.

4.3

Matrix Math

Mathcad can perform virtually any of the math operations often required when working with matrices, with or without units. Some of these are shown in Fig. 4.17. Try reproducing it, keeping in mind:

1. Use the Vector and Matrix Toolbar as necessary.

2. Each expression shown is standard matrix math, except the one showing $A + 3$. Strictly speaking you cannot add a scalar to a matrix, but Mathcad allows you to (it's intended to be a useful option), and the result is as shown: the scalar is added to all elements in the matrix. (Do not confuse this with adding, say, $3I$, where I is the identity matrix!)

3. The worksheet shows how to solve a matrix equation $Ax = b$, for x.

In addition to the operations shown, Mathcad has many other functions (some of which we'll cover in the exercises at the end of this section), such as:

1. Stacking matrices, and selecting a submatrix from a matrix.

2. Counting the number of columns or rows of a matrix.

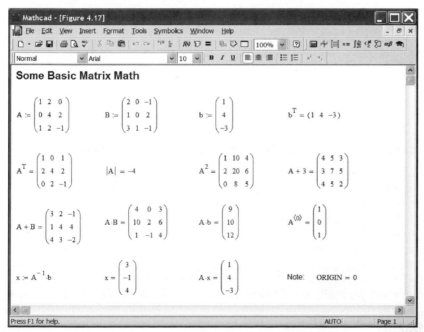

**Figure 4.17:
Some basic matrix
(or array) math**

3. Finding the maximum and minimum values in a matrix.

4. Using more advanced concepts, such as matrix rank; eigenvalues and eigenvectors (see the next section); and Cholesky, LU, and QR decomposition.

We demonstrate basic operations in matrix (or array) math by doing an example.

Example 4.6 (Example 3.5 *revisited*): Matrix Math with Units: Currents in a DC Network Given the network shown in Fig. 4.18, find the currents I_1, I_2, and I_3.

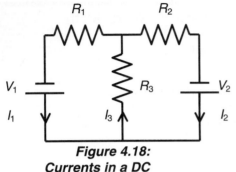

Figure 4.18: Currents in a DC circuit

$R_1 = 2\ \Omega$
$R_2 = 4\ \Omega$
$R_3 = 3\ \Omega$
$V_1 = 12\ \text{V}$
$V_2 = 9\ \text{V}$

Applying Kirchoff's law and summing voltage drops across each battery, we obtain three equations for the three unknowns, I_1, I_2, and I_3:

$$I_1 + I_2 - I_3 = 0$$
$$R_1 I_1 + R_3 I_3 = V_1$$
$$R_2 I_2 + R_3 I_3 = V_2$$

We can rewrite these as:

$$AI = b$$

where

$$A = \begin{bmatrix} 1 & 1 & -1 \\ R_1 & 0 & R_3 \\ 0 & R_2 & R_3 \end{bmatrix}, I = \begin{bmatrix} I_1 \\ I_2 \\ I_3 \end{bmatrix} \text{ and } b = \begin{bmatrix} 0 \\ V_1 \\ V \end{bmatrix}$$

The worksheet of Fig. 4.19 shows the solution; it's a good idea to try to reproduce it. Some important comments can be made:

1. *Be careful with your subscripts!* The resistor and voltages all have literal subscripts. The only place we use array subscripts is in the final row, to evaluate the three currents.

2. We changed the value of ORIGIN to unity; if we didn't we would have to evaluate I_0, I_1, and I_2 instead of I_1, I_2, and I_3 (which would not be a notation consistent with the problem).

3. *Be careful about defining away built-in units!* We committed a sin here: we defined a matrix as A in a worksheet to do with an electrical problem in which we really want to also keep Mathcad's default use of A for amps! On this occasion we got away with it (we can still evaluate the currents and get A for amps), but we were taking a risk. It would have been better to define, say, matrix M instead.

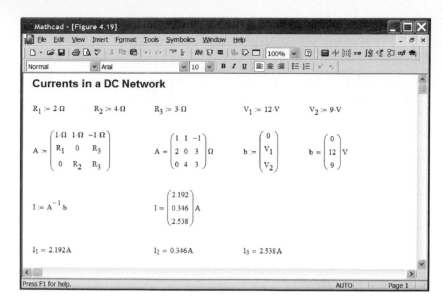

Figure 4.19:
Currents in a DC
network

Exercises

For these exercises, use:

$$A = \begin{bmatrix} 1 & 2 & 4 \\ 1 & 0 & -1 \\ 2 & -2 & 1 \end{bmatrix}, b = \begin{bmatrix} 1 \\ 2 \\ -1 \end{bmatrix}, \text{ and } B = \begin{bmatrix} 1 & 0 & 2 \\ 0 & 3 & 1 \\ 1 & 2 & 1 \\ 1 & 1 & 2 \end{bmatrix}.$$

4.20 Write down your prediction of which of the following products can and cannot be executed; then compute all of them to check your predictions: $A \cdot b$, $b \cdot A$, $A \cdot B$, $B \cdot A$, $B \cdot A \cdot b$, $b \cdot A \cdot B$, A^2, b^2, and B^2.

4.21 Write down your prediction of the size (e.g., 3×2) of the following products; then compute all of them to check your predictions: $A \cdot A^T$, $A^T \cdot A$, $b \cdot b^T$, $b^T \cdot b$, $B \cdot B^T$, and $B^T \cdot B$.

4.22 By evaluating $(A^T)^{-1}$ and $(A^{-1})^T$, verify that taking the transpose and the inverse of a matrix are commutative.

4.23 If $A \cdot x = b$, find x.

4.24 If $A \cdot x = B^T$, find x.

4.25 Evaluate the result of creating a new matrix by stacking A and b^T and subtracting from this matrix B. (Look under Mathcad's Help for "stack.")

4.26 Evaluate the result and create a new matrix by augmenting A with b, taking the transpose of this new matrix, and subtracting from this matrix B. (Look under Mathcad's Help for "augment.")

4.4

An important topic in linear algebra is finding the eigenvalues and eigenvectors of a square matrix. This math comes up in a number of engineering problems (e.g., multiple-degree-of-freedom vibrations and stress analysis). We will illustrate Mathcad's power in this topic through an example.

**Eigenvalues
and Eigenvectors**

**Example 4.7: Eigenvalues and Eigenvectors: Principal Stresses
and Axes** The stresses measured at a point in a beam are:

$$\tau = \begin{bmatrix} \sigma_{xx} & \tau_{xy} & \tau_{xz} \\ \tau_{yx} & \sigma_{yy} & \tau_{yz} \\ \tau_{zx} & \tau_{zy} & \sigma_{zz} \end{bmatrix} = \begin{bmatrix} 3 & -2 & 5 \\ -2 & -4 & 1 \\ 5 & 1 & 2 \end{bmatrix} \cdot \text{MPa}$$

where the σ represents normal stresses and the τ stands for shear stresses. Find the maximum and minimum normal stresses, in the material, and the axes on which these occur.

The idea behind principal stresses is that, no matter what stresses may be applied to a material, an orientation can always be chosen such that the stresses on the surfaces are purely normal (tension or compression); these are the principal stresses, and their orientation is the principal axes. For example, if the two-dimensional normal and shear stresses shown in Fig. 4.20(a) are applied to a material element, the purely normal stresses shown in Fig. 4.20(b) are also present. In this example, the normal stresses are 5 MPa tension and 1 MPa compression; the principal stresses are 7 MPa tension and 3 MPa compression. In general, to find the principal stresses we can use a two- or three-dimensional Mohr's circle method (as appropriate). Alternatively, it turns out that if we represent the stresses as a 2×2 or 3×3 matrix (as appropriate), the eigenvalues of the matrix will be the principal stresses, and the eigenvectors will be their orientation with respect to the original axes. We use Mathcad for the latter method. The worksheet shown in Fig. 4.21 contains the solution to our current problem. Our comments on this are:

1. *Be careful with your subscripts*! We use both literal and array subscripts here. All subscripts are literal, except subscript-subscripts. You might want to try using array subscripts throughout instead to see what happens.

2. We introduce two new built-in Mathcad functions:
   a. eigenvals( ): This must be typed lowercase, and its argument *must* be a square matrix. It generates the eigenvalues for the matrix (2 for 2×2 matrix, 3 for a 3×3 matrix, etc.). In our case the matrix is symmetric (it is always symmetric for a stress problem) so it turns out the eigenvalues will always be real, but in general they could be complex.
   b. eigenvec(,): This must also be typed lowercase, and its arguments *must* be a square matrix and an eigenvalue of that matrix. It generates the eigenvector for that eigenvalue (and it is normalized–i.e., its magnitude

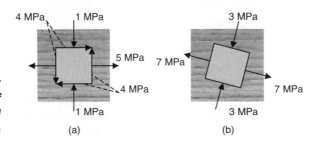

*Figure 4.20:
Example of
two-dimensional
stresses*

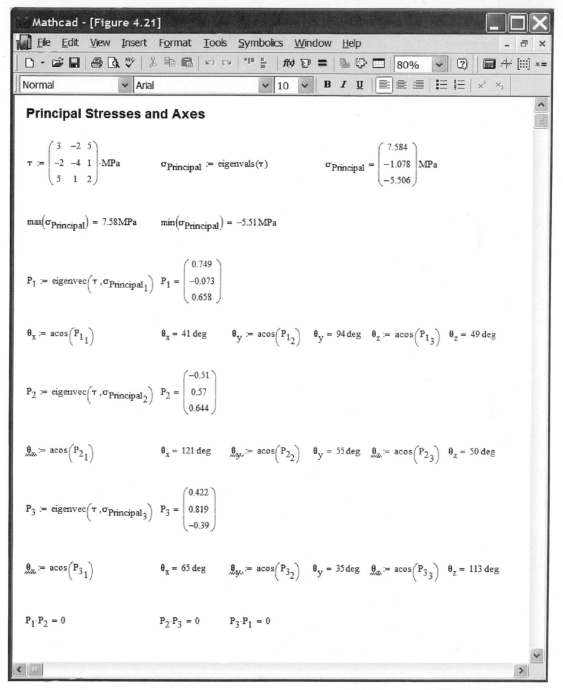

Figure 4.21:
*Eigenvalues and
eigenvectors: principal
stresses and axes*

is unity). For our symmetric matrix it turns out that the eigenvectors are orthogonal to one another, as we demonstrate by computing the dot products in the last line. There is also an eigenvecs( ) function, which takes only the matrix as argument and gives all eigenvectors as a single matrix.

3. The θ's are the direction cosines for each principal axis, that is, the angles each axis makes with the original coordinates; we inserted units of degs.

Mathcad has many built-in functions for manipulating vectors and matrices. We will explore a number of them in the exercises that follow.

Additional Exercises

4.27 The amount of paint (in gallons) of black, red, green, and white needed to paint the living room, bedroom, bathroom, and dining room of an apartment is given by the following table:

	Black	Red	Green	White
Living	1	2	0	2
Bed	0	4	2	1
Bath	2	0	3	1
Dining	1	3	1	3

The paints cost ($/gal):

	Black	Red	Green	White
Cost	27	22	29	18

Find the total cost of paint per room.

4.28 For Exercise 4.27, a different paint supplier was used, leading to paint costs of:

	Living	Bed	Bath	Dining
Cost ($)	104	144	152	164

Find the cost ($/gal) of each color of paint.

4.29 Find the loop currents I_1, I_2, and I_3, in the circuit shown, and find the heat generated. Summing voltage drops around each loop, we obtain three equations:

$$R_1I_1 + R_4(I_1 - I_2) - V_1 = 0$$
$$R_2I_2 + R_4(I_2 - I_1) + V_2 = 0$$
$$R_3I_3 + V_3 - V_2 = 0$$

Rewritten, these become $Ax = b$, where

$$A = \begin{bmatrix} (R_1 + R_4) & -R_4 & 0 \\ -R_4 & (R_2 + R_4) & 0 \\ 0 & 0 & R_3 \end{bmatrix}, b = \begin{bmatrix} V_1 \\ -V_2 \\ (V_2 - V_3) \end{bmatrix}, \text{and } x = \begin{bmatrix} I_1 \\ I_2 \\ I_3 \end{bmatrix}.$$

The heat generated in a resistor R by current I is RI^2. Notes: Set ORIGIN to 1; use literal subscripts for everything except the I's; and use units. *This is the same problem as Exercise 3.15.*

4.30 Repeat Exercise 4.29 if V_1 is removed and replaced with a connector (i.e., a zero-resistance connection). *This is the same problem as Exercise 3.16.*

4.31 Repeat Exercise 4.29 if V_2 is removed and replaced with a resistor R_5 = 10 Ω. *This is the same problem as Exercise 3.17.*

4.32 Find the loop currents I_1, I_2, I_3, and I_4, in the circuit shown, and find the heat generated. Note: Summing voltage drops around each loop, we obtain four equations:

$$R_1 I_1 + R_5(I_1 - I_2) - V_1 = 0$$
$$R_2 I_2 + R_5(I_2 - I_1) + V_2 = 0$$
$$R_3 I_3 + R_6(I_3 - I_4) - V_2 = 0$$
$$R_4 I_4 + R_6(I_4 - I_3) + V_3 = 0$$

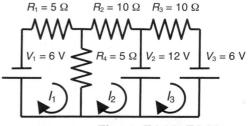

**Figure E4.29, E4.30,
E4.31**

The heat generated in a resistor R by current I is RI^2.
Notes: Refer to Exercise 4.29 to see how to restructure the equations into a matrix format $Ax = b$; set ORIGIN to 1; use literal subscripts for everything except the I's; and use units. *This is the same problem as Exercise 3.43.*

4.33 Repeat Exercise 4.32 if V_1 is removed and replaced with a connector (i.e., a zero-resistance connection). *This is the same problem as Exercise 3.44.*

4.34 Repeat Exercise 4.32 if V_2 is removed and replaced with a resistor R_7 = 10 Ω. *This is the same problem as Exercise 3.45.*

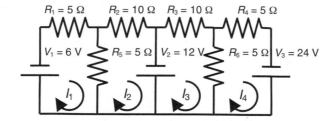

**Figure E4.32, E4.33,
E4.34**

4.35 Cramer's rule is a method used to solve a matrix equation of the form $A \cdot x = b$. Instead of solving by writing $x = A^{-1} \cdot b$, we solve by computing $x_i = \dfrac{D_i}{D} = \dfrac{|A_i|}{|A|}$, where i = 1, 2, 3, etc. In this method, $D = |A|$ is the determinant of matrix A, and $D_i = |A_i|$ is the determinant of matrix A_i, which is matrix A with the ith column replaced with b (Phew!). Solve Exercise 4.23 using this method.

4.36 Two masses are constrained to vibrate between two walls by three springs, as shown. There are an infinite number of ways we could have these two masses vibrate. For example, we could hold the one on the left in place while we force the second one an inch to the right—when we let go of both, we will get a pretty complicated motion; another type of motion would occur if we simply whacked one or both masses! It turns out that all such possible complicated free vibration motions can be built from a combination of two "pure" vibrations: the fundamental modes, in which each mass vibrates in a perfect sine wave. In one mode, both masses vibrate in sync; in the second they vibrate exactly out of sync. The natural frequencies of these motions are given by $f = \dfrac{1}{2\pi}\sqrt{-\lambda}$, where λ are the eigenvalues of the following matrix:

$$
M = \begin{bmatrix} -\left(\dfrac{k_1 + k_2}{m_1}\right) & \dfrac{k_2}{m_1} \\[2ex] \dfrac{k_2}{m_2} & -\left(\dfrac{k_2 + k_3}{m_2}\right) \end{bmatrix}
$$

Find the two natural frequencies (Hz). Note: Use literal subscripts for the masses and springs; look up *vectorize* in Help to see how to compute the f's in one equation.

4.37 In Exercise 4.36 if spring k_3 is removed, find the two natural frequencies.

4.38 In Exercise 4.36 suppose spring k_2 is made very stiff (e.g., $k_2 = 10^6$ lbf/in). Find the two natural frequencies and compare their values to $\dfrac{1}{2\pi}\sqrt{\dfrac{k_1 + k_3}{m_1 + m_2}}$. Can you explain this?

Figure E4.36, E4.37, E4.38

$k_1 = 20$ lbf/in $k_2 = 15$ lbf/in $k_3 = 20$ lbf/in

$m_1 = 10$ lb $m_2 = 5$ lb

Symbolic Math and Calculus

Mathcad has some wonderful capabilities in symbolic math (by symbolic math we mean math operations that lead to an algebraic rather than a numerical result). You can use Mathcad two ways to evaluate symbolically: the Symbolics menu and the Symbolic Keyword Toolbar.

Is It Live or Is It Not Live . . .?

Using the Symbolics Menu

You can use the Symbolics menu shown in Fig. 5.1. *These transformations are not live, but are one-time calculations.* From Fig. 5.1, we can see that we can do lots of basic things with an expression, such as simplifying and expanding, factoring, and solving for a variable, as well as more sophisticated operations such as differentiating and integrating, and performing Laplace or Fourier transforms.

In this method, what happens depends on the settings you have set up in the Evaluation Style window, shown in Fig. 5.2 with its default settings. The "Show evaluation steps" radio buttons allow

Figure 5.1: Using the Symbolics menu

you to control where the result will be placed—below or to the right of the original; the Show Comments checkbox controls whether or not Mathcad automatically inserts explanatory text between the original expression and the result; and Evaluate In Place, if checked, means the result replaces the original expression.

Figure 5.3, which you should try to reproduce, shows some examples using this method. We can make some comments here:

1. For this worksheet, we set the evaluation style so that comments are shown, and the evaluation steps are inserted horizontally.

2. After typing each expression, the *entire* expression was selected before selecting the relevant command from the Symbolics menu. (Remember that you can select the entire expression by clicking anywhere in it and repeatedly pressing **Spacebar**. (Note that if you press **Spacebar** too many times—after you've already selected the expression—you will cycle back down to selecting only one variable.) If you don't do this, and instead select only part of an expression or only a variable, for example, x or θ, Mathcad will attempt to perform the operation just on what you've selected! Sometimes this is what you want to do; for example, you might want to simplify or factor only part of a larger expression.

3. a. Examples 1 and 2 illustrate Mathcad's ability to demonstrate some well-known identities. For each use menu item *Symbolics . . . Simplify*.
  b. Example 3 shows that "simplify" can mean something different to you and me than what it means to Mathcad.
  c. Examples 4 and 5 together show that "simplify" and "expand" can often be synonyms (use *Symbolics . . . Factor* and *Symbolics . . . Expand* for factoring and expanding, respectively).
  d. Examples 6 and 7 show that "expand" and "factor" are the opposite of one another.
  e. Example 8 is a second example of factoring.

4. Although this doesn't occur in Fig. 5.3, there is no guarantee that use of the Symbolics menu will succeed; if it fails, Mathcad usually gives back the original expression. The method will obviously fail if there is no symbolic result; it will also fail for expressions that are particularly difficult to manipulate. You should consider Mathcad's symbolic math capability to be very good, but not at genius level!

5. All these results are static—if you change any of the original expressions, the resulting expression does *not* change! Try editing any of the expressions to verify this.

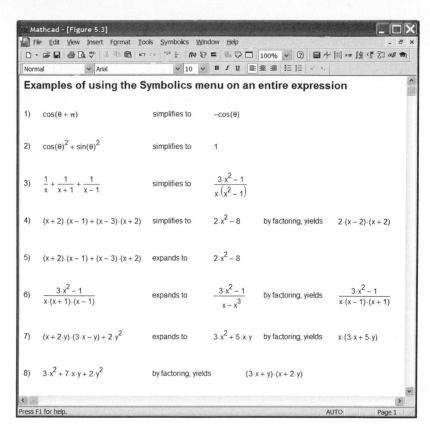

Examples of using the Symbolics menu on an entire expression

1) $\cos(\theta + \pi)$ simplifies to $-\cos(\theta)$

2) $\cos(\theta)^2 + \sin(\theta)^2$ simplifies to 1

3) $\dfrac{1}{x} + \dfrac{1}{x+1} + \dfrac{1}{x-1}$ simplifies to $\dfrac{3 \cdot x^2 - 1}{x \cdot (x^2 - 1)}$

4) $(x+2) \cdot (x-1) + (x-3) \cdot (x+2)$ simplifies to $2 \cdot x^2 - 8$ by factoring, yields $2 \cdot (x-2) \cdot (x+2)$

5) $(x+2) \cdot (x-1) + (x-3) \cdot (x+2)$ expands to $2 \cdot x^2 - 8$

6) $\dfrac{3 \cdot x^2 - 1}{x \cdot (x+1) \cdot (x-1)}$ expands to $-\dfrac{3 \cdot x^2 - 1}{x - x^3}$ by factoring, yields $\dfrac{3 \cdot x^2 - 1}{x \cdot (x-1) \cdot (x+1)}$

7) $(x + 2 \cdot y) \cdot (3 \cdot x - y) + 2 \cdot y^2$ expands to $3 \cdot x^2 + 5 \cdot x \cdot y$ by factoring, yields $x \cdot (3 \cdot x + 5 \cdot y)$

8) $3 \cdot x^2 + 7 \cdot x \cdot y + 2 \cdot y^2$ by factoring, yields $(3 \cdot x + y) \cdot (x + 2 \cdot y)$

Figure 5.3:
Using the Symbolics menu on an entire expression

Figure 5.4, which you should next reproduce, shows additional symbolic operations using the Symbolics menu. Comments for this worksheet are:

1. For this worksheet, we again set the evaluation style so that comments are shown, and the evaluation steps are inserted horizontally.

2. After typing each expression, *any one instance of the variable* we're interested in must be selected before using the relevant command from the Symbolics menu. For example, in Example 1 there are three instances of x in the original expression; you can select any one of them before selecting *Symbolics . . . Collect.* If you mistakenly select the entire expression when you should be selecting just a variable, either you will get the uninformative error message shown in Fig. 5.5, or the symbolic operation will be grayed-out and unavailable. The error message in Fig. 5.5 might lead you to conclude that you can't perform the symbolic math you requested, when in fact the error was just in not selecting a variable!

a. Examples 1 and 2 illustrate the importance of selecting a variable before using *Symbolics . . . Collect.* In Example 1 we selected an x, and in Example 2 we selected a y. In each case, Mathcad rearranged the expression and collected terms in descending powers of x or y, respectively—for example, in Example 1, x^2 has a coefficient 3, x has a coefficient $(2y - 2)$, and there is a "constant" $2y^2$ (compare this to Example 2, where we collected the same expression in powers of y).

b. Example 3 also shows the importance of variable selection. In the first case, we selected an x, in the second a y. We then used *Symbolics . . . Polynomial Coefficients* to generate an array listing the coefficients of x^0, x^1, and x^2 (and then y^0, y^1, and y^2). Note that using *Collect* and *Polynomial Coefficients* gives essentially the same results, just in differing formats: *Collect* produces an expression in powers of the selected variable, with the coefficients deduced; *Polynomial Coefficients* gives the same coefficients, but in a matrix format.

c. Example 4 shows what happens if you select an x or a y in an expression without an equals sign, and use *Symbolics . . . Variable . . . Solve*. With *Solve*, Mathcad assumes the expression is set to zero (in other words, it finds the roots of the expression in the selected variable).

d. Example 5 uses *Solve* for an equation rather than an expression; note that you *must* use the Boolean equals (type **Ctrl + =** or use the ▣ **=** icon on the Boolean Toolbar). In this case there is one solution, but in general there could obviously be several.

e. Example 6 shows the effect of using *Symbolics . . . Variable . . . Convert to Partial Fraction* on any one of the x's in the expression.

f. Example 7 shows Mathcad's ability to generate a Taylor or Laurent series (we'll see some examples of this in the exercises at the end of the chapter); use *Symbolics . . . Variable . . . Expand to Series*.

g. Example 8 shows that Mathcad is pretty good at calculus; we differentiate and then integrate to show these operations are the inverse of one another (use *Symbolics . . . Variable . . . Differentiate* or *Integrate*).

h. Example 9 shows a substitution. We replaced x with $\theta + \frac{1}{2}\pi$ throughout the expression; however, there is a trick here—you must have *previously* copied or cut the expression you wish to insert (in the example this is $\theta + \frac{1}{2}\pi$) so that Mathcad knows what it is to replace the current variable with (in this example, x). Use *Symbolics . . . Variable . . . Substitute*.

i. Example 10 shows Mathcad performing a Laplace transform on variable θ, and then its inverse transform; note that the transform is to s–space and the inverse transform is to t–space, even though we started with θ–space.

3. There is no guarantee that use of the Symbolics menu will succeed; in that case, Mathcad usually gives back the original expression.

4. All these results are static—if you change any of the original expressions, the resulting expression does *not* change! Try editing any of the expressions to verify this.

Using the Symbolic Keyword Toolbar

The Symbolic Keyword Toolbar, shown in Fig. 5.6, can be accessed by using *View . . . Toolbars . . . Symbolic*, or by clicking on the 🎓 icon on the Math Toolbar.) This method is *live*: if you change the original expression, the symbolic math result immediately updates (which is *not* true with the Symbolics menu). The icons on the toolbar are again fairly obvious, but we will illustrate with some examples.

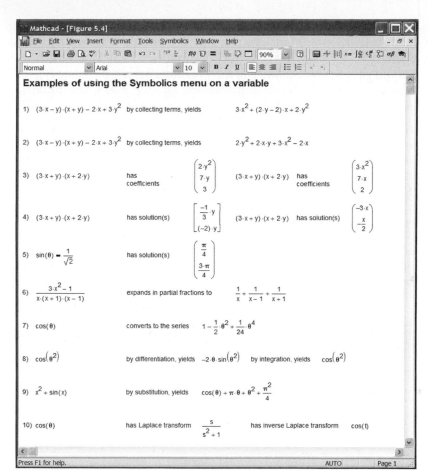

Figure 5.4:
Using the Symbolics
menu on a variable in
an expression

Note that although we generally must use the toolbar, there are two keystroke combinations available as well:

1. You can use **Ctrl + .**, which immediately symbolically evaluates a selected expression (it generates the $\rightarrow$ icon shown on the toolbar in Fig. 5.6). Note that in most cases . . . .

2. You can use **Ctrl + Shift + .**, which allows you to type one of the expressions in Fig. 5.6, such as **expand**, in a placeholder, to have more control over which symbolic procedure gets executed (it generates the $\bullet \rightarrow$ icon shown on the toolbar in Fig. 5.6).

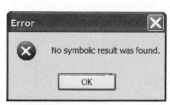

Figure 5.5:
The symbolic result error
message

$\rightarrow$	$\bullet \rightarrow$	Modifiers	float	rectangular
assume	solve	simplify	substitute	factor
expand	coeffs	collect	series	parfrac
fourier	laplace	ztrans	invfourier	invlaplace
invztrans	$M^T \rightarrow$	$M^{-1} \rightarrow$	$|M| \rightarrow$	explicit
combine	confrac	rewrite		

Figure 5.6:
The Symbolic Keyword Toolbar

We will illustrate how to use of most of the icons in the Symbolic Keyword Toolbar. For example, Fig. 5.7 shows many of the same operations we used in the Symbolics menu of Figs. 5.3 and 5.4. You should practice some basic symbolic math by reproducing Fig. 5.7. Many examples are self-explanatory, but we make the following comments:

1. For each example shown, after typing the expression, click on the appropriate icon in the Symbolic Keyword Toolbar or, to use only the keyboard, type **Ctrl + Shift + .**, then in the placeholder type the appropriate operation (e.g., type **simplify** for Examples 1 and 2).

2. Sometimes using the icons introduces a placeholder that you may or may not want to use. For example, clicking the simplify icon usually generates not only the word "simplify" but also a placeholder; you can type something in the placeholder, or delete it. Note that some operations *require* something in the placeholder or placeholders. For example, the substitute icon requires two entries—the variable to be replaced and the variable or expression to replace it with.

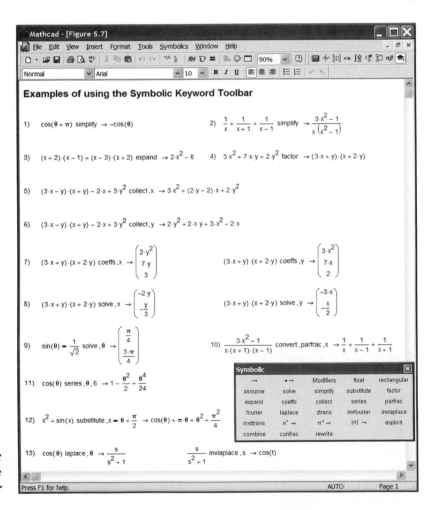

3. All of these computations are live! For example, try editing Example 1 to so that it is $\cos(\theta + \frac{\pi}{2})$, and you'll immediately get $-\sin(\theta)$ instead of $-\cos(\theta)$ as the result.

Which one of these two methods—the Symbolics menu or the Symbolic Keyword Toolbar—depends to some degree on your personal preference. We note when each is usually the best approach:

1. Use the Symbolics menu when you're writing a technical report because the *appearance* of the symbolic math is pretty much how you'd type or write a report; you don't have to worry about Mathcad's strange (to non-Mathcad readers) symbolic arrows and keywords.

2. Use the Symbolic Keyword Toolbar (or keystrokes) for pretty much everything else. I almost always prefer this method because it's live. After all, one of the best things about Mathcad is the ability to do what-ifs by changing initial values and getting immediately new results!

In the rest of this chapter, we will use exclusively the Symbolic Keyword Toolbar approach.

5.2

As we have seen in Figs. 5.3, 5.4, and 5.7, Mathcad is convenient for performing much of your symbolic math. There are many ways to tweak Mathcad's symbolic math. In the following examples, we look at some of these: we learn how to control the appearance of symbolic math; how to use variables that have values, but still do symbolic math; the effect of the decimal point on symbolic evaluations; how to string together several symbolic math operations; and how to use *explicit* to further control output. There are many other ways to manipulate Mathcad's symbolic math—see Mathcad's *Help* on symbolic math.

Some Symbolic Math "Bells and Whistles"

Example 5.1 (Examples 3.5 and 4.6 *revisited*): Symbolic Math— Matrix Math with Values: Currents in a DC Network Given the network shown in Fig. 5.8, find the currents I_1, I_2, and I_3.

$R_1 = 2\ \Omega$
$R_2 = 4\ \Omega$
$R_3 = 3\ \Omega$
$V_1 = 12\ V$
$V_2 = 9\ V$

Referring to Example 4.6, the solution is:

$I = A^{-1}b$

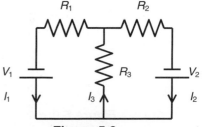

Figure 5.8: Currents in a DC circuit

where

$$I = \begin{bmatrix} I_1 \\ I_2 \\ I_3 \end{bmatrix} \quad \text{with} \quad A = \begin{bmatrix} 1 & 1 & -1 \\ R_1 & 0 & R_3 \\ 0 & R_2 & R_3 \end{bmatrix}, \quad and \ b = \begin{bmatrix} 0 \\ V_1 \\ V_2 \end{bmatrix}$$

The worksheet of Fig. 5.9 shows the solution; it's a good idea to try to reproduce it. First, we repeat some important comments about arrays from Chapter 4:

1. *Be careful with your subscripts!* The resistor and voltages all have literal subscripts (e.g., for R_1 type **R.1**). The only place we use array subscripts is when we evaluate the three currents (e.g., for I_1 type **I[1]**).

2. We changed the value of ORIGIN to unity; if we didn't we would have to evaluate I_0, I_1, and I_2 instead of I_1, I_2, and I_3 (which would not be a notation consistent with the problem).

3. *Be careful about defining away built-in units!* We committed a sin here: we defined a matrix as A, making Mathcad's understanding of A ambiguous (Is it A for the matrix or A for amps?). It would have been better to define, say, matrix M instead.

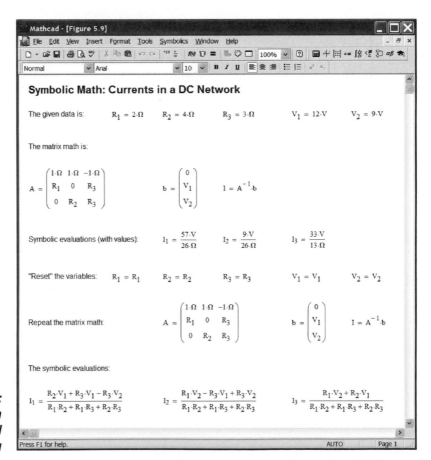

Figure 5.9:
Using symbolic math
with values and
formatting

In addition to the above, we used some tricks to make the worksheet look more professional:

1. Under *Tools ... Worksheet Options ... Display,* we changed the *Definition* and *Symbolic Evaluation* views to *Equal.* This makes all equations display the standard equals sign (=), making the worksheet look like regular math (which is good), but it hides from us (until you click on an individual expression) the logic behind the worksheet (which is bad). For example, we no longer see when we're assigning and when we're evaluating.

2. All the expressions use the assignment equals (type **:**), except the six expressions in which we evaluate the currents I_1, I_2, and I_3. For the first set of these expressions, we typed **Ctrl + .** (we could also have clicked on the $\rightarrow$ icon on Symbolic Keyword Toolbar). For the second set, we typed **Ctrl + Shift + .**, and in the placeholder typed **simplify** (we could also have clicked on the simplify icon on Symbolic Keyword Toolbar). We then right-clicked on each of the equations to access some convenient display options, as shown in Fig. 5.10. In the worksheet of Fig. 5.9, we checked *Hide keywords* so that, on exiting the equation, the keyword "simplify" is not displayed.

3. In Fig. 5.9 we symbolically evaluated the three currents twice, to illustrate an important point: *when symbolically evaluating an expression that includes variables that have assigned values, Mathcad will use those values!* The first half of the worksheet in Fig. 5.9 shows this. Sometimes this is the desired result, but usually we want a truly symbolic answer. There *is* a trick to get Mathcad to give the true symbolic result, even in this case: *redefine each variable that has an assigned value to be equal to itself.* Then, when you define a new variable in terms of these variables, symbolic evaluation will be truly symbolic. The last half of the worksheet in Fig. 5.9 shows this. There is one more subtlety in doing symbolic math with variables that include values: *if your values are not integers, you will get symbolic results with many significant figures!* This is illustrated in Fig. 5.11, which shows what happens if we change R_1 to $2 \cdot \Omega$ instead of

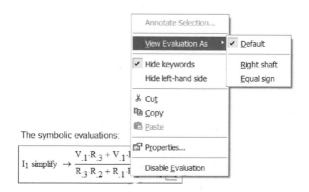

Figure 5.10: Symbolic display options

$R_1 = 2 \cdot \Omega$ ... leads to ... $I_1 \rightarrow \dfrac{57}{26 \cdot \Omega} \cdot V$

$R_1 = 2.0 \cdot \Omega$ ... leads to ... $I_1 \rightarrow \dfrac{2.1923076923076923076}{\Omega} \cdot V$

Figure 5.11: Effect of a decimal point

2·Ω (That's right—just add a decimal point!). To reduce the number of significant figures displayed, either restructure the input value to make it integers, or use the float icon on the Symbolic Keyword Toolbar. An example of "restructuring" an input would be to replace, say 0.25, with $\frac{1}{4}$.

Example 5.2: Symbolic Math: Pressure Drop Through a Venturi Meter A Venturi meter is an inexpensive device for measuring the flow rate Q in a pipe. It consists of a reduction of diameter from pipe diameter D_1 to D_2, as shown in Fig. 5.12. At this "throat," the flow speed increases from V_1 to V_2; therefore, the fluid pressure drops (subject to certain assumptions). This pressure drop, Δp, can be measured using a manometer, leading to a manometer height reading Δh. Hence, the manometer reading Δh indicates the flow rate Q. The equations for the Venturi and manometer are

$$\frac{\Delta p}{\rho} = \frac{V_2^2}{2} - \frac{V_1^2}{2}$$

$$V_1 = \frac{Q}{\frac{\pi}{4}D_1^2} \quad V_2 = \frac{Q}{\frac{\pi}{4}D_2^2}$$

$$\Delta p = (\rho_m - \rho)g\Delta h$$

Solve the above equations to find Δh in terms of Q, D_1, D_2, ρ_m (the manometer fluid density), ρ (the pipe fluid density), and g. Then find Δh (in) for the following data: $Q = 200$ gal/min, $D_1 = 3$ in, $D_2 = 2$ in, $\rho_m = 13500$ kg/m³, and $\rho = 1000$ kg/m³.

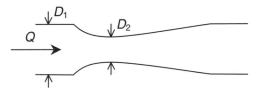

**Figure 5.12:
The Venturi meter**

The worksheet of Fig. 5.13 shows the solution; try and reproduce it. Some hints at developing it are:

1. We use literal subscripts throughout.

2. We (intentionally) used mixed unit systems (e.g., gal, kg) to show that Mathcad can still handle the problem.

3. Under *Tools . . . Worksheet Options . . . Display*, we once again changed the *Definition* and *Symbolic Evaluation* views to *Equal* to make all equations display the standard equals sign (=). See the comments on Example 5.1 for the pros and cons of doing this.

4. After specifying the values of Q, D_1, D_2, ρ_m, and ρ, we immediately redefined them to be equal to themselves, so that the subsequent symbolic math would actually be symbolic.

5. All the expressions use the assignment equals (type **:**), *except* the equations in which we first define and symbolically evaluate Δh. These two equations are shown in all their gory detail in Fig. 5.14, and are discussed below.

6. The equation in which we define Δh is a compound one, in which we take the first equation, substitute for V_1, V_2, and Δp, and then solve for Δh! To generate it:

a. We set up the first of the equations (using the Boolean equals).

b. Then we type **Ctrl + Shift + .** and in the placeholder type **substitute,** (which triggers another placeholder in which we type the equation for V_1; we could also click on the substitute icon on Symbolic Keyword Toolbar and obtain two placeholders, one for each side of the equation).

c. To get the next line, we can either type **Ctrl + Shift + .** and **substitute,** (or click on the substitute icon again), and type the equation for V_2.

d. We continue typing **Ctrl + Shift + .** and **substitute,** (or click on the substitute icon one more time) to make the substitution for Δp, *except we need to be a bit careful*—make sure the entire previous "substitute" procedure is selected first, or you may insert the new line between the previous two (it's a quirk of Mathcad). If this happens you can just delete the line and start over.

e. We use a final line to solve for Δh (type **Ctrl + Shift + .** and **solve,** or click on the solve icon).

7. Now that we have symbolically solved for Δh, we can see what happens when we substitute values for all the variables, without performing all the canceling, as shown in the next equation in Figs. 5.13 and 5.14. To do

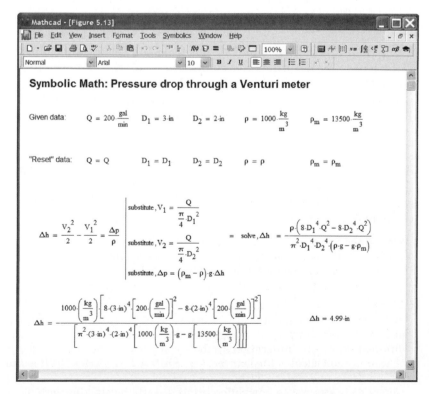

Figure 5.13: Multiple symbolic operations; pressure drop through a Venturi meter

$$\Delta h = \frac{{V_2}^2}{2} - \frac{{V_1}^2}{2} = \frac{\Delta p}{\rho} \left| \begin{array}{l} \text{substitute}, V_1 = \dfrac{Q}{\frac{\pi}{4} \cdot {D_1}^2} \\[6pt] \text{substitute}, V_2 = \dfrac{Q}{\frac{\pi}{4} \cdot {D_2}^2} \\[6pt] \text{substitute}, \Delta p = (\rho_m - \rho) \cdot g \cdot \Delta h \end{array} \right. \quad \rightarrow \; \text{solve}, \Delta h \rightarrow -\frac{\rho \cdot \left(8 \cdot {D_1}^4 \cdot Q^2 - 8 \cdot {D_2}^4 \cdot Q^2 \right)}{\pi^2 \cdot {D_1}^4 \cdot {D_2}^4 \cdot (\rho \cdot g - g \cdot \rho_m)}$$

**Figure 5.14:
Compound symbolic
math, and use of
explicit**

$$\Delta h \text{ explicit}, \Delta h, Q, D_1, D_2, \rho, \rho_m = -\frac{1000 \cdot \left(\frac{kg}{m^3} \right) \left[8 \cdot (3 \cdot in)^4 \left[200 \cdot \left(\frac{gal}{min} \right) \right]^2 - 8 \cdot (2 \cdot in)^4 \left[200 \cdot \left(\frac{gal}{min} \right) \right]^2 \right]}{\left[\pi^2 \cdot (3 \cdot in)^4 \cdot (2 \cdot in)^4 \left[1000 \cdot \left(\frac{kg}{m^3} \right) \cdot g - g \cdot \left[13500 \cdot \left(\frac{kg}{m^3} \right) \right] \right] \right]}$$

this, type **Ctrl + Shift + .** and **explicit,** (or click on the explicit icon), and then type the variable list shown in Fig. 5.14. What *explicit* does is allow you to control which variables have their values substituted, and which not. If we just symbolically evaluate Δh we get the symbolic answer shown in the compound expression (check this by typing **D Ctrl + g h Ctrl + .**); using *explicit* with one or more variables substitutes values for just those variables. This is meant as a convenience, since you may want a reader of your worksheet to see intermediate calculations. There are a couple of Mathcad quirks here: we needed to include Δh as one of our variables, and include at least one of $Q, D_1, D_2, \rho,$ and ρ_m; otherwise, we just get Δh back (experiment with your worksheet by eliminating some variables in the expression). Notice also that Mathcad does *not* substitute the value of g, even if you include it in your explicit list. (You'd have to define g at the top of the worksheet, even though Mathcad knows what it is, to get it to work in the explicit list!)

For this example, we could have solved the set of equations for Δh by hand, but it's still convenient to have Mathcad do the grunt work! The only down side to this is that the final formula for Δh could be made more compact—we could ask Mathcad to do this for us, but it's sometimes quicker to do it manually.

5.3

**Calculus—
Symbolic and
Numeric**

Mathcad is an excellent tool for performing differentiation and integration—both symbolic and numeric. Figure 5.15 shows some examples of this (you can practice calculus by reproducing it). First, we opened the Calculus Toolbar. Note the following:

1. We defined variables $x, y, a,$ and b with specific values, and immediately redefined them to be equal to themselves (remember, this is so that subsequent symbolic computations *will* be symbolic).

2. We defined two functions $f(x)$ and $g(x,y)$.

3. We performed several examples of calculus:

    a. Example 1 shows Mathcad's ability to perform basic differentiation and integration, evaluated using symbolic evaluation. To generate the differentiation and integration symbols just click on the $\frac{d}{dx}$, $\frac{d^n}{dx^n}$, $\int_a^b$, or $\int$ icon in the Calculus Toolbar (or type **Shift + /**, or **Ctrl + Shift + /** to differentiate, and **Shift + 7** (the **&** key) or **Ctrl + i)** to integrate; what you *cannot* do is attempt to type differentials directly, by starting with d!

b. Example 2 shows the exact same operations, but this time the regular evaluation equals sign is used. In this case, Mathcad inserts known values of variables. For the differentiations it uses $x = 3$, and for the definite integral it integrates and *then* substitutes the limits $x = a$ and $x = b$. Note that not even Mathcad can give you the value of an indefinite integral!

c. Example 3 shows that Mathcad can perform partial differentiation. To do this, simply perform a differentiation for the variable of interest as previously described (in the example, first x then y)! Mathcad will display the equation as an ordinary derivative; to make it look like a partial, simply right-click on any part of the derivative, and in the menu that appears select *View Derivative As . . . Partial Derivative.*

d. Example 4 shows Mathcad's ability to integrate with variable limits; the second expression shows a double integral (simply use the $\int_\cdot^\cdot$ icon or type **Shift + 7** twice)—we simplified the result after seeing that without doing so led to a complicated expression (use the simplify icon on the Symbolic Keyword Toolbar or type **Ctrl + Shift + .** and then in the placeholder type **simplify**).

e. In Example 5 we see that Mathcad can handle infinite integrals (to get ∞ use the ∞ icon or type **Ctrl + Shift + z**). The first equation leads to a complicated result because, depending on the value of n, different results will occur (try typing **n:2** or **n:–2** above the example); in the second, we inserted a constraint on n (use the assume icon on the Symbolic Keyword Toolbar, or type **Ctrl + Shift + .** and then in the placeholder type **assume,**).

4. We could have performed all the math shown in Fig. 5.15 using the Symbolics menu instead of the Symbolic Keyword Toolbar.

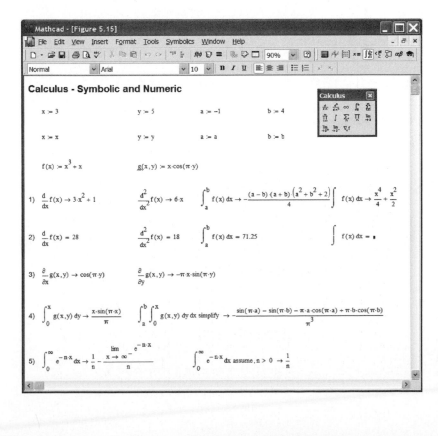

Figure 5.15:
Examples of symbolic and numeric calculus

Example 5.3: Symbolic Integration: Volume and Surface Area of a Spherical Shape

Prove by integrating that the volume and surface area of a sphere of radius R are $4\pi R^2/3$ and $4\pi R^2$, respectively. Then find θ (see Fig. 5.16) such that the volume and surface area of a segment of the sphere are $\pi R^2/2$ and $\pi R^2/2$, respectively. The integrals for the volume and surface area of a sphere are:

$$V = \int_0^{2\pi} \int_0^{\pi} \int_0^R r^2 \sin(\theta)\,dr\,d\theta\,d\phi$$

$$A = \int_0^{2\pi} \int_0^{\pi} R^2 \sin(\theta)\,d\theta\,d\phi$$

The solution to this problem is shown in Fig. 5.17. After all the work in this chapter, you should have little problem reproducing it. The following notes apply:

1. Make sure the integration limits are consistent with the integration variables: we work *outward* from the integrand to find the matching pairs.

2. We reproduce the first pair of integrals and right-click on the evaluation and symbolic equals signs of each to get a menu for displaying them as regular equals signs (just because we could!).

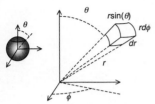

***Figure 5.16:
Geometry for
sphere calculation***

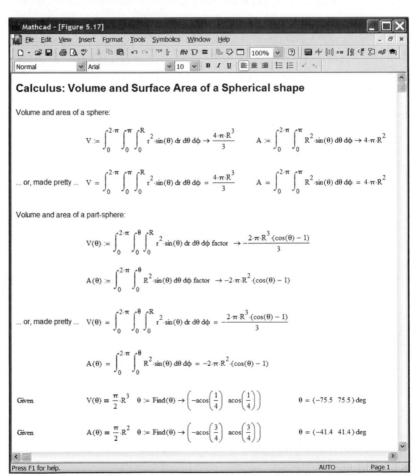

***Figure 5.17:
Volume and surface
area of a sphere***

3. For the second part of the problem, we define two functions $V(\theta)$ and $A(\theta)$ to be the results of symbolic calculation, as shown (and again reproduced "prettier" versions). Note that in each we integrate from $\theta = 0$ to $\theta = \theta$!

4. We use Mathcad's Given . . . Find feature twice to find the correct values of θ. Note that we are asking Mathcad to symbolically use the Given . . . Find involving finding an appropriate upper limit of an integral. Also, we don't need an initial guess value for the θ's, even though we end up evaluating them, because we used Given . . . Find symbolically—a neat trick!

In this chapter we have barely scratched the surface on the symbolic capabilities of Mathcad. We haven't done much more than mention some topics, such as use of the Taylor series and Laplace transforms. We'll see Taylor series in some exercises at the end of the chapter. You should also be aware of Mathcad's fairly comprehensive help on symbolic math—click on *Help* and search for symbolic math to get the window shown in Fig. 5.18.

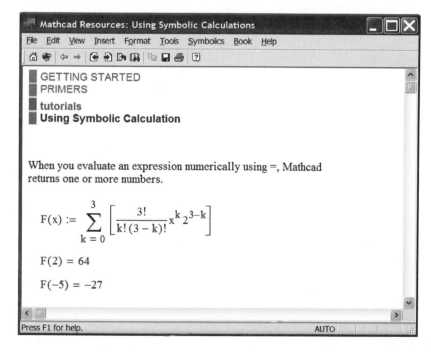

*Figure 5.18:
Additional help on
symbolic math*

Exercises

Note: Be aware that with symbolic math you may encounter problems if you solve one problem that involves a variable (say, θ), and then use the same variable in a later problem in the same worksheet. If you get some strange results, this may be the reason; if so, simply solve those problems on separate worksheets.

5.1 Expand (a) $\cos(\alpha + \beta) + \cos(\alpha - \beta)$, b) $\sin(\alpha + \beta) + \sin(\alpha - \beta)$.

5.2 Expand (a) $\cos(2\theta)$, (b) $\sin((2\theta)$.

5.3 Factor $x^3 + x^2 - 4x - 4$.

5.4 Convert to partial fractions $\dfrac{x^2 + 12x + 60}{x^3 + 3x^2 - 10x - 24}$.

5.5 Take the derivative of, and simplify, the expression in Exercise 5.4. Then integrate the result.

5.6 Consider the functions $u(x) = x^2 + x$ and $v(x) = e^x$. Use them to demonstrate the method of integration by parts, which is

$$\int u(x) \cdot \frac{d}{dx} v(x)dx = u(x) \cdot v(x) - \int v(x) \cdot \frac{d}{dx} u(x)dx.$$

Note: After defining $u(x)$ and $v(x)$, symbolically compute the left and right sides of the equation, and compare.

5.7 Evaluate symbolically $\displaystyle\sum_{m=0}^{\infty} x^m$.

5.8 Evaluate symbolically $\displaystyle\sum_{m=0}^{\infty} \frac{x^m}{m!}$.

5.9 Evaluate symbolically $\displaystyle\sum_{m=0}^{\infty} \frac{(-1)^m \theta^{2m}}{(2m)!}$.

5.10 Evaluate symbolically $\displaystyle\sum_{m=0}^{\infty} \frac{(-1)^m \theta^{2m+1}}{(2m+1)!}$.

5.11 Evaluate symbolically $\displaystyle\lim_{n\to\infty} \sqrt[n]{\frac{1}{n}}$ ..

5.12 Symbolically evaluate $\displaystyle\lim_{N\to\infty}\left(\sum_{k=1}^{N} \frac{1}{k} - \ln(N)\right)$ and try to find out what the result means. (Hint: Check out "Euler" online!)

5.13 Solve symbolically $x^2 - 8x + 20 = 0$.

5.14 Symbolically evaluate $(a + b{\cdot}i){\cdot}(c + d{\cdot}i)$, using the complex keyword. (Note: To create an i that you intend to be $\sqrt{-1}$, you *cannot* just type **i**, but must type **1i** (no space)!)

5.15 Symbolically evaluate $\displaystyle\int \frac{1}{x^2 + a^2}\,dx$.

5.16 Symbolically evaluate $\dfrac{d}{d\theta} \tan^{-1}(\theta)$. (Note: $\tan^{-1}$ is really *atan*!)

5.17 Symbolically evaluate $\dfrac{d}{dx}(\sinh(x) + \cosh(x))$.

5.18 The temperature of a cooling body is given by $T = T_0 e^{-\frac{t}{\tau}}$, where T_0 is the initial temperature, and τ is a time constant. Solve for t.

5.19 Your bank balance F is given by:

$$F = P \cdot \left(1 + \frac{i}{k}\right)^n$$

where P is the principal (initial investment), i is the annual interest rate, k is the number of compounding periods a year, and n is the total number of periods. Solve for i, simplify, and factor to obtain the simplest possible expression for i.

5.20 The APR (annual percentage rate) is the actual interest received at the end of the year, based on the advertised interest rate being compounded daily, monthly, and so on. If the original deposit is P, the annual interest rate is i, and the number of compounding periods in the year is n, the balance after a year is given by $F = P \cdot \left(1 + \frac{i}{n}\right)^n$. Show that for in the limit $n \to \infty$, this becomes exponential, $F = P \cdot e^i$.

5.21 Symbolically solve $\tan(\theta) = 1$.

5.22 Symbolically solve $\sin(\theta) \cos(\theta) = \frac{1}{2}$.

5.23 A supersonic airplane flies at Mach number M. The half-angle of the nose of the airplane is θ. An important factor is the pressure ration across the oblique shock, $p_r = p/p_{atm}$, where the pressures are as shown:

$$p_r = 1 + \frac{2k}{k+1}(M^2 \sin^2(\beta) - 1)$$

Symbolically solve for b, and evaluate for $M = 2$ and $p_r = 2.5$ ($k = 1.4$).

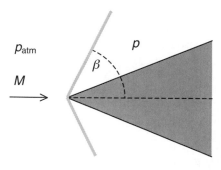

Figure E5.23

Figure E5.24

where $k = 1.4$ is the ratio of air specific heats, and β is the angle of the oblique shock.

5.24 The initial voltage across the capacitor shown is V_0. Prove that after the switch is closed, the total heat generated in the resistor is the same as the initial energy stored in the capacitor $\left(\frac{1}{2} V_0^2 C\right)$. Note: The heat generated in the resistor is given by $Q = \int_0^\infty \frac{V_0^2}{R} e^{-\frac{2t}{RC}} dt$; you will need to evaluate this, assuming $R > 0$ and $C > 0$!

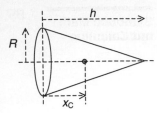

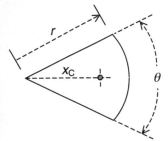

Figure E5.25

Figure E5.26

5.25 Prove that the volume of the right circular cone shown is $V = \frac{1}{3}\pi R^2 h$, and that the center of gravity is at $x_c = \frac{h}{4}$. Use the following formulas:

$$V = \int_0^h \pi R^2 \left(1 - \frac{x}{h}\right)^2 dx \text{ and } x_c = \frac{1}{V}\int_0^h \pi R^2 x \left(1 - \frac{x}{h}\right)^2 dx.$$

5.26 Prove that the area of the "pizza slice" shown is $A = \frac{1}{2}\theta r^2$, and that the center of gravity is at $x_c = \frac{4r}{3\theta}\sin\left(\frac{1}{2}\theta\right)$. Use the following formulas:

$$A = \int_{\frac{\theta}{2}}^{\frac{\theta}{2}}\int_0^r r \, dr \, d\theta \text{ and } x_c = \frac{1}{A}\int_{\frac{\theta}{2}}^{\frac{\theta}{2}}\int_0^r r^2 \cos(\theta) dr \, d\theta.$$

5.27 Find the loop currents $I_1, I_2,$ and $I_3,$ in the circuit shown, by symbolically solving the following set of equations:

$$R_1 I_1 + R_4(I_1 - I_2) - V_1 = 0$$
$$R_2 I_2 + R_4(I_2 - I_1) + V_2 = 0$$
$$R_3 I_3 + V_3 - V_2 = 0$$

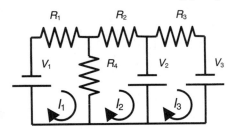

Figure E5.27, E5.28

5.28 Rewritten, the equations in Exercise 5.27 become $Ax = b$, where

$$A = \begin{bmatrix} (R_1 + R_4) & -R_4 & 0 \\ -R_4 & (R_2 + R_4) & 0 \\ 0 & 0 & R_3 \end{bmatrix}, b = \begin{bmatrix} V_1 \\ -V_2 \\ (V_2 - V_3) \end{bmatrix}.$$

and $x = A^{-1}b$ represents the unknown currents. Solve for x. Note: Use literal subscripts throughout.

5.29 Find the loop currents $I_1, I_2, I_3,$ and $I_4,$ in the circuit shown. Note: Summing voltage drops around each loop, we obtain three equations

$$R_1 I_1 + R_5(I_1 - I_2) - V_1 = 0$$
$$R_2 I_2 + R_5(I_2 - I_1) + V_2 = 0$$
$$R_3 I_3 + R_6(I_3 - I_4) - V_2 = 0$$
$$R_4 I_4 + R_6(I_4 - I_3) + V_3 = 0$$

5.30 Repeat Exercise 5.29 if V_1 and V_3 are removed and replaced with connectors (i.e., zero-resistance connections).

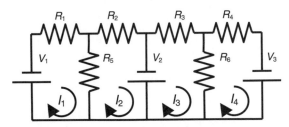

Figure E5.29, E5.30

5.31 Find the first 16 terms of the Taylor series of $\sin(\theta^2)$ and of $2\theta \cos(\theta^2)$ around $\theta = 0$, and by doing so demonstrate that one is the derivative of the other.

5.32 Find the Taylor series of $\cos(2\pi x)$ about $x = 0$ three different times: for five terms, for seven terms, and for nine terms, and plot all three as well as $\cos(2\pi x)$ against x, from $x = 0$ to $x = 0.5$ (see Appendix: Graphing for help in setting up the graph). Discuss the meaning of your results.

5.33 Find the first three terms of the Taylor series of $e^{-\frac{x}{5}}$ about $x = 0$ and about $x = 20$. Plot both with $e^{-\frac{x}{5}}$ against x, from $x = 0$ to $x = 15$ (see Appendix: Graphing for help in setting up the graph). Discuss the meaning of your results.

Numerically Solving Differential Equations

Mathcad has quite a number of built-in methods for numerically solving differential equations. Some of these are:

1. *Odesolve*: This method allows you to solve ordinary differential equations in a compact, easy-to-understand format (it uses a *Given . . . Odesolve* layout similar to the *Given . . . Find* method we used for solving algebraic equations). In the background, it uses one of the methods listed in item 2. If you wish, you can go directly to pages 108, 116, or 122 for help on this.

2. *rkfixed*, *Rkadapt*, *or Bulstoer*: These are essentially built-in fourth-order Runge-Kutta (RK4) methods for solving ODEs. If you wish, you can go directly to pages 113, 118, or 121 for help on this.

3. *Pdesolve*: This method allows you to solve partial differential equations (PDEs) in a compact, easy-to-understand format. It is for use with hyperbolic and parabolic systems. We will not discuss this method in this book (see Mathcad's Help).

We will practice each of these methods in this chapter, as well as an approach where we build our own set of equations for analyzing such problems.

6.1

Numerically Solving a Single First-Order ODE

We first look at methods for solving a first-order ODE, with initial condition, of the standard form

$$\frac{dy}{dx} = f(x, y), \qquad y(x_0) = y_0$$

The methods we will apply are:
1. The Euler method (or Euler-Cauchy method).
2. The *improved* Euler method (or Heun's method).
3. The *Odesolve* method built into Mathcad.
4. The *rkfixed* method, also built in to Mathcad.

The Euler Method for a Single ODE

We already introduced the Euler method in Example 4.3. The idea behind the Euler method is very simple: starting with the differential equation, we wish to solve

$$\frac{dy}{dx} = f(x, y), \qquad y(x_0) = y_0$$

we realize that graphically the derivative $\frac{dy}{dx}$ is the slope of the solution curve $y(x)$ (i.e., the solution we wish to obtain). If we are at some point (x_n, y_n) on the curve, we can follow the tangent at that point, as an approximation to actually moving along the curve itself, to find a new value for y, y_{n+1}, corresponding to a new x, x_{n+1}, as shown in Fig. 6.1. We have

$$\frac{dy}{dx} = \frac{y_{n+1} - y_n}{x_{n+1} - x_n}$$

If we choose a *step size* $h = x_{n+1} - x_n$, then the above equation can be combined with the differential equation to give

$$\frac{dy}{dx} = \frac{y_{n+1} - y_n}{h} = f(x_n, y_n)$$

or

$$y_{n+1} = y_n + hf(x_n, y_n)$$

with

$$x_{n+1} = x_n + h$$

In these equations y_{n+1} now represents our best effort to find the next point on the solution curve. From Fig. 6.1, we see that y_{n+1} is *not* on the solution curve, but close to it; if we make the triangle much smaller, by making the step size h smaller, then y_{n+1} will be even closer to the desired solution. This is the Euler method approach. We can repeatedly use the two Euler iteration equations to start at (x_0, y_0) and obtain (x_1, y_1), then (x_2, y_2), (x_3, y_3), and so on; we don't end up with an equation for the solution, but a set of numbers.

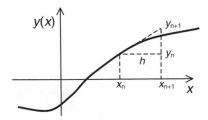

Figure 6.1:
The Euler method

We can make a few specific comments:

1. It is very easy to set up, making it an attractive approach.

2. Because we built it ourselves, it turns out we can use units, a big plus!

3. It is not very accurate: following the tangent to a curve at each point, in an attempt to follow the curve, is pretty crude!

4. If we make the step size h smaller, the accuracy of the method will generally increase, but obviously you then need more steps to achieve the solution. It turns out if you use too many steps (if h is extremely small) the accuracy of the results can decrease because although each small step is very accurate, you will now need so many of them that round-off errors can build up.

5. As with any numerical method, you are not guaranteed to get a solution, or one that is very accurate!

Let's illustrate the method with an example.

Example 6.1: The Euler Method for a First-Order ODE A tank contains water at an initial depth $y_0 = 10$ ft. The area of the cross section of the tank is $A = 25$ ft². A hole of area $a = 1$ in² appears at the bottom of the tank. (See Fig. 6.2.) Using a 10-point Euler method, estimate the water depth after $t_{end} = 30$ min, compute the error compared to the exact solution, and plot the Euler and exact results.

It turns out that the ODE describing this is

$$\frac{dy}{dt} = -\frac{a}{A}\sqrt{2gy} \qquad y(0) = y_0$$

and the exact solution is

$$y_{Exact}(t) = \left(\sqrt{y_0} - \frac{a}{A}\sqrt{\frac{g}{2}}\,t\right)^2$$

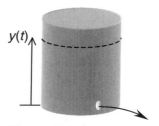

Figure 6.2: Flow from a tank

The solution is shown in Fig. 6.3, which you should try to reproduce. Our comments are:

1. Be careful with subscripts: all of them, except t_{end} and y_{Exact}, are *array* subscripts!

2. The differential equation is just for display; it is not used in any computation.

3. Note that we discussed the Euler method in terms of $y(x)$, but in this problem we used $y(t)$.

4. The general form of the function is $f(t,y)$, but note that in this particular problem it is a function of y only.

5. Once we define the range $n = 0, 1, 2, \ldots, 10$, *any equation involving n is immediately computed for all n values*, which is exactly what we want to do with the two iteration formulas.

6. We are able to use units, a big benefit (especially with British units); we also insert units at various evaluations (e.g., the default unit for time is seconds, so evaluating h gives 180 s; we then inserted min at the units placeholder).

7. We evaluate t_N to verify that we do have the correct end time when $n = N$. We then evaluate y_N, and $y_{Exact}(t_N)$—remember that y and t are columns of values, and $y_{Exact}(t)$ is a continuous function that we evaluate at t_N.

8. We leave it to you to figure out how to generate the nicely formatted graph. See Appendix: Graphing if (when!) you get stuck!

9. Choosing $N = 10$ is a pretty crude way to slice the time, showing an error after 30 min of about 18%; try changing N to 20, 50, 100 etc., to see that the accuracy does in fact improve—and see the graph get very busy!

10. We present the exact solution so we can see how good the Euler method is, but of course you wouldn't bother with the Euler method if you could get the exact solution!

The Euler method is most useful when you have a complicated ODE, and also when you have inputs that involve such special terms as Heaviside step functions (see Example 6.4).

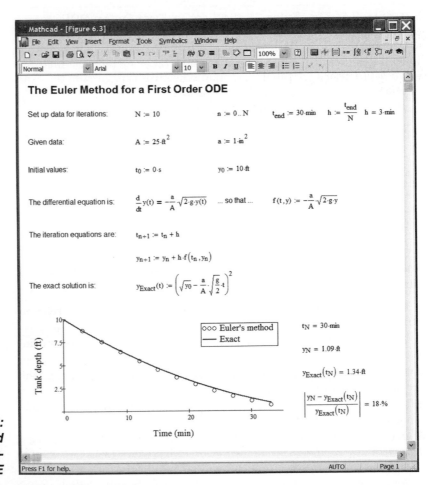

*Figure 6.3:
The Euler method
applied to a first-
order ODE*

The Euler method resulted in a simple iteration formula:

$$y_{n+1} = y_n + hf(x_n, y_n)$$

In this equation, we estimate a later value of y, y_{n+1}, from an earlier value, y_n, by following the tangent to the curve at (x_n, y_n), computed from $f(x_n, y_n)$. Suppose instead for each n we compute

$$y_{int_{n+1}} = y_n + hf(x_n, y_n)$$

and

$$y_{n+1} = y_n + \frac{1}{2} h[f(x_n, y_n) + f(x_{n+1}, y_{int_{n+1}})]$$

In these equations, we first obtain an estimate of the next y value as in the Euler method ($y_{int_{n+1}}$), but then we use this as a first *prediction* for obtaining a better estimate for the next y value! In the second equation, we follow a tangent that is the average of the slope of the curve at the beginning (x_n) and (approximately) at the end (x_{n+1}); in essence, our first prediction helps us to *correct* the tangent, to make it follow the curve more accurately. This method is an example of a *predictor-corrector* method.

To use this method in Mathcad, we need to combine the equations to get

$$y_{n+1} = y_n + \tfrac{1}{2}h[f(x_n, y_n) + f(x_{n+1}, y_n + h(f(x_n, y_n)))]$$

Yes, this is a bit messy! However, it's much more accurate than the Euler method for the same number of points, and it is not too difficult to use in Mathcad.

Example 6.2: The Improved Euler Method for a First-Order ODE Recompute the solution to Example 6.1 using the improved Euler method.

The solution is shown in Fig 6.4, which you should try to reproduce. Our comments are:

1. The 10 comments we made above for the Euler method apply here!

2. Notice that Figs. 6.3 and 6.4 show that the modified Euler method is significantly more accurate than the Euler method for the same number of points. It is slightly harder to set up, but then it is more effective.

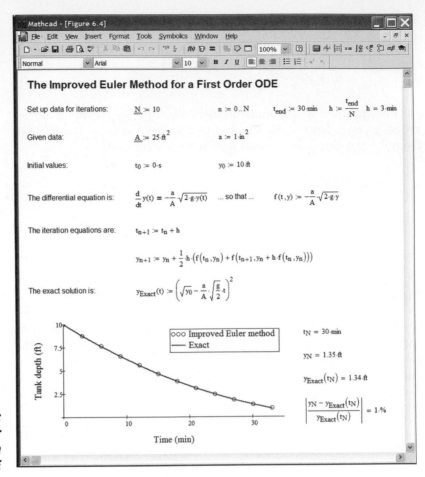

*Figure 6.4:
The improved Euler
method applied to a
first-order ODE*

The Euler and improved Euler methods are both quick and simple "homemade" ODE solvers. Their chief advantages are:

1. It's clear what you're computing.

2. We can solve problems that include units.

We now turn to some built-in ODE solvers.

Mathcad's Odesolve Method for a Single First-Order ODE

This is a very powerful and user friendly feature of Mathcad. We illustrate its use by revisiting our now familiar leaking tank problem.

Example 6.3: The Odesolve Method for a First-Order ODE

Recompute the solutions for Examples 6.1 and 6.2 using the improved Euler method.

The solution is shown in Fig 6.5, which you should try to reproduce. Our comments on using Odesolve (in this example, and also in general) are:

1. You cannot use units with this built-in method, so for a problem that has units, they must be removed. Hence, we take each piece of data and redefine it to be itself divided by its own standard units. For example, we divide t_{end} (= 30 min) by s to get a dimensionless t_{end} whose *value* (1.8×10^3) is in s; similarly, we start with $A = 25$ ft², and end up with $A = 2.323$ (meaning 2.323 m²). If we *started* with a problem with all quantities in standard units (in SI these would be quantities such as m for length, s for time, N for force, etc.), we could just type in those values; otherwise we need to perform this tedious procedure using Mathcad or by performing the conversions by hand. In any event, *all variables must be dimensionless, with values corresponding to standard units.*

2. For the *Odesolve* you *must* start with the math region *Given*.

3. The ODE to be solved *must* be typed with the following rules:

　a. The dependent variable must be typed *with its independent variable argument* wherever it appears in the differential equation. In Fig. 6.5 this means we must type $y(t)$ not y in two places.

　b. The equation *must* be typed using the Boolean equals.

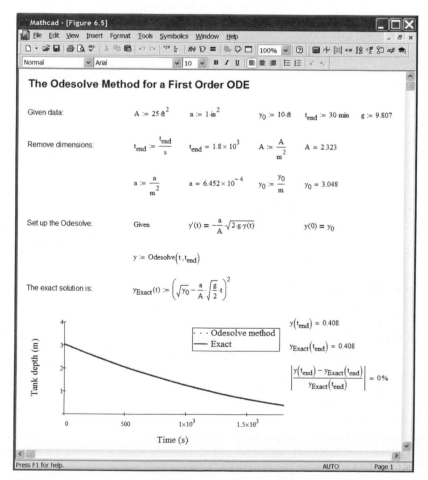

Figure 6.5:
The Odesolve method applied to a first-order ODE

c. The differential *must* be created using the ✳ icon in the Calculus Toolbar (or type **Shift + /**). However, instead of the differential, you may use the more compact notation involving the prime symbol, shown in Fig. 6.6. To get this, type ` (on the tilde key, ~) or type **Ctrl + F7** (Yes, it's a bit odd!); you *cannot* type the apostrophe, ' (try it to see what happens!).

$$\frac{d}{dt}y(t) = -\frac{a}{A}\sqrt{2 \cdot g \cdot y(t)} \qquad y'(t) = -\frac{a}{A}\sqrt{2 \cdot g \cdot y(t)}$$

Using the ✳ icon or typing **Shift + /**   Typing ` or **Ctrl + F7**

d. The initial condition *must* be typed using the Boolean equals.

e. The Odesolve statement is used to define y as shown. The two arguments that must be used are the independent variable (in this case t), and the end point (in this case t_{end}); an optional final argument (not used here) is the number of steps used in the solution—the default is 1000. Note that we type **y**, not **y(t)**, even though the statement *will* generate a new function, $y(t)$!

4. The Odesolve function uses a Runge-Kutta numerical method to solve the differential equation. The RK4 method is a more sophisticated method than the improved Euler method discussed above. Note that both the Euler and improved Euler methods generate a set of values for y (they generate a column of values), but the Odesolve goes one better: it uses the column of y values generated by the RK4 method, and then creates a continuous function $y(t)$ from these values! The default RK4 method used by Odesolve is fixed (i.e., the step size is constant), but you can right-click on *Odesolve* and select *Adaptive* to make Mathcad automatically use smaller step sizes in regions where the function is changing rapidly—a pretty neat trick!

5. *All* subscripts in Fig. 6.5 are literal subscripts.

6. Note that the Euler method (Fig. 6.1) had mediocre accuracy, the improved Euler method (Fig. 6.4) was better, but the Odesolve method (Fig. 6.5) is terrific!

In the next Example, we use *Odesolve* with a special function, the Heaviside step function.

Example 6.4: The Odesolve Method and the Heaviside Step Function—Lag in a Control System The response of a control system can often be described by

$$\frac{dy}{dt} = A\frac{F(t)}{\tau} - \frac{y}{\tau}$$

where $A = 2$ is the system gain, $\tau = 0.75$ is the system time constant, and $F(t)$ is the input to the system. Plot the response $y(t)$ of the system from $t = 0 - 5$, and find the peak response, for the input shown in Fig. 6.7 ($y(0) = 0$).

For the function $F(t)$ we need to develop a piecewise continuous function. There are a number of ways to do this; for example, we could use *if* functions (built into Mathcad—see Fig. 6.10 below). Here we use the "Heaviside step function", defined as

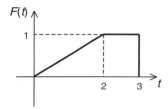

$$\Phi(\alpha) = \begin{cases} 0 & if \quad \alpha < 0 \\ 1 & if \quad \alpha \geq 0 \end{cases}$$

Figure 6.7:
The input function f(t)

We start by defining $F(t) = \dfrac{1}{2}t$ up to the point $t = 2$. How do we next proceed? The easiest approach is to visualize "killing" the function by subtracting $\dfrac{1}{2}t$, and also adding a new function $F(t) = 1$, but only after $t = 2$! A little thought shows that we must multiply these new functions by $\Phi(t-2)$, which is zero until we reach $t = 2$. This is illustrated in Fig. 6.8.

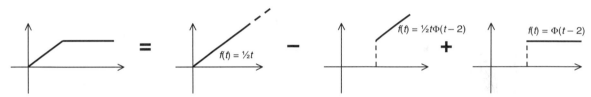

Figure 6.8:
Building part of F(t)

Finally, we must "kill" $F(t) = 1$ at $t = 3$ by subtracting 1 multiplied by $(t-3)$! We end up with

$$F(t) = \tfrac{1}{2}t + \Phi(t-2)\cdot(1 - \tfrac{1}{2}t) - (t-3)$$

Now we have the input function defined, the Odesolve method becomes easy. You should try reproducing the worksheet shown in Fig. 6.9. The following comments (some of which are worth repeating from Example 6.3) apply:

1. To get the Heaviside step function type **F** and then immediately type **Ctrl + G**, or use the *Greek Symbol Toolbar*. The most common error with using this function is to use a multiply (*) between the Φ and the parentheses: $\Phi\cdot(t-2)$ makes no more sense than would, say, $\sin\cdot(\theta)$!

2. Remember that for the Odesolve you must start with the math region Given.

3. The ODE to be solved must be typed with the rules already outlined in Example 6.3:

    a. The dependent variable must be typed *with its independent variable argument* wherever it appears in the differential equation. In Fig. 6.9 this means we must type $y(t)$ not y in two places.

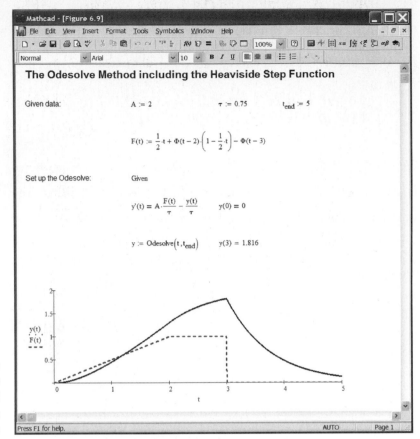

b. The equation *must* be typed using the Boolean equals.

c. The prime differential in the ODE equation *must* be created by typing ` (on the tilde key, ~), or typing **Ctrl + F7**; you *cannot* type the apostrophe, '. Alternatively, you may create a regular-looking derivative using the $\frac{d}{dx}$ icon in the *Calculus Toolbar* (or by typing **Shift + /**). These alternatives were shown in Fig. 6.6.

d. The initial condition must be typed using the Boolean equals.

e. The Odesolve statement is used to define *y* as shown; the two arguments that must be used are the independent variable (in this case *t*), and the end point (in this case t_{end}). Note that we type **y**, not **y(t)**, even though the statement *will* generate a new function, $y(t)$!

4. Although the Odesolve function uses a Runge-Kutta numerical method in the background, you can right-click on the *Odesolve* and select *Adaptive* to make Mathcad automatically use smaller step sizes in regions where the function is changing rapidly.

5. The peak is clearly at $t = 3$, so we evaluate $y(t)$ there.

6. We leave it to you to figure out how to generate the nicely formatted graph (see Appendix: Graphing for help). Note that the function $F(t)$ may appear to have a slight dent as it drops from down to zero—it's just an artifact of the plotting.

Note that Mathcad has at least one more way to create a piecewise continuous function: the *if* function. This is defined and used as follows: *if(cond,x,y)* generates result *x* if *cond* is true; otherwise, it generates result *y*. This is illustrated in Fig. 6.10, where we show

$$F(t) := \frac{1}{2} \cdot t + \Phi(t-2) \cdot \left(1 - \frac{1}{2} \cdot t\right) - \Phi(t-3) \qquad F(t) := \frac{1}{2} \cdot t + if\left[t \geq 2, \left(1 - \frac{1}{2} \cdot t\right), 0\right] - if(t \geq 3, 1, 0)$$

Using the Heaviside step function

Using the if function

Fig. 6.10: Creating piecewise continuous functions

how to create the function $F(t)$ using the Heaviside and if functions (the $\geq$ symbol is obtained by typing **Ctrl + 0** or from the *Boolean Toolbar*).

A final comment on Example 6.4: this problem can be solved analytically in various ways. For example, we could solve the ODE from $t = 0$ to $t = 2$ for $F(t) = \frac{1}{2} t$; then we could use the final value of y, $y(2)$, as the initial condition for a second solution, during which $F(t) = 1$; and finally we could use the final value of this solution, $y(3)$, as the initial value for a third solution, during which $F(t) = 0$. This is a lot of work! A slightly easier approach might be to use Mathcad's built-in Laplace transform capabilities.

We mentioned that *Odesolve* uses a built-in Runge-Kutta differential equation solver, which simply makes the worksheet look a bit friendlier. We can also access this RK4 method directly, as we discuss next.

Mathcad's rkfixed Method for a Single First-Order ODE

This is the routine that Odesolve uses; using it directly, as we'll do now, is not as pretty, but you're able to exercise more control over what happens. We illustrate with an example.

Example 6.5: The rkfixed Method for a First-Order ODE Recompute the solution to Example 6.4 using the *rkfixed* method.

The solution is shown in Fig 6.11, which you should try to reproduce. Our comments on using *rkfixed* for a single ODE (in this example, and also in general) are:

1. The *rkfixed* method requires you to set up the differential equation in standard form

$$\frac{dy}{dt} = f(t, y)$$

In the worksheet of Fig. 6.11, we used the original equation in Example 6.4

$$\frac{dy}{dt} = A\frac{F(t)}{\tau} - \frac{y}{\tau}$$

so

$$f(t, y) = A\frac{F(t)}{\tau} - \frac{y}{\tau}$$

2. The *rkfixed* method will solve the ODE at discrete points; in other words, it will not produce a continuous function but will instead create a column of y values for the solution. In the worksheet $N = 500$ is the number of points.

3. You *must* create a new function $D(t,y)$ that is equal to the right-hand side of the standard ODE, that is, $f(t,y)$. Note that we now have two functions that are computationally identical! We did this for one reason only: we wanted to be consistent with the notation used in Mathcad's Help on *rkfixed*, which uses this D notation (D is supposed to stand, I suppose, for "differential"). You may use only $f(t,y)$ or only $D(t,y)$ throughout the worksheet if you wish.

4. You *must* create a new variable (in our worksheet S for "solution") and set it equal to the *rkfixed* function. The function is used in the following way:

rkfixed(initial value of y, initial value of t, final value of t, number of points, D)

In the current worksheet, we have *rkfixed*(0, 0, t_{end}, N, D), based on our data. Note that D must be entered without its arguments (i.e., D not $D(t, y)$)!

5. The variable S is the solution to the differential equation. It is a matrix of values, *not* a function! The first column of the matrix is the 501 values of the independent variable, in our case t (t_0 through t_N), generated by *rkfixed*; the second column is the corresponding 501 values of the dependent variable, in our case y (y_0 through y_N), also generated by *rkfixed* (you can verify all this by typing **S:** at the end of the worksheet).

6. To extract the values of the dependent (y) and independent (t) variables from S, we need to use the *Matrix Column* icon ( ) on the Vector and Matrix Toolbar, or type **Ctrl + 6**. Note: In Fig. 6.11 the default value of *ORIGIN* is zero, so the first column of S is <0>; if you have *ORIGIN* set to unity, the first column will be <1>!

7. Because y is an array, we can ask for its max value, as shown.

8. To plot the result with the original function $F(t)$, we need to define a range variable n as shown in Fig. 6.11. This is because y and t are both arrays (column vectors); they *can* be plotted directly against one another, but if we want to plot a *function* ($F(t)$) versus array t, we must generate an array by computing $F(t_n)$. For more discussion on plotting arrays, see Appendix: Graphing.

Note that we used *rkfixed* here, but we could also have used other built-in functions, such *Rkadapt* (which changes the step size in t to optimize the accuracy). For more on this, search for *ODE solvers* in Mathcad's Help to get the window shown in Fig. 6.12.

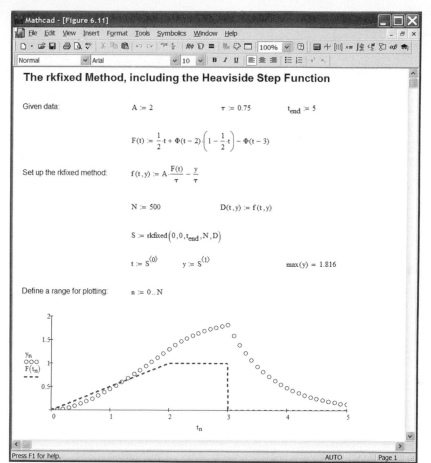

*Figure 6.11:
The rkfixed method
applied to a first-order
ODE*

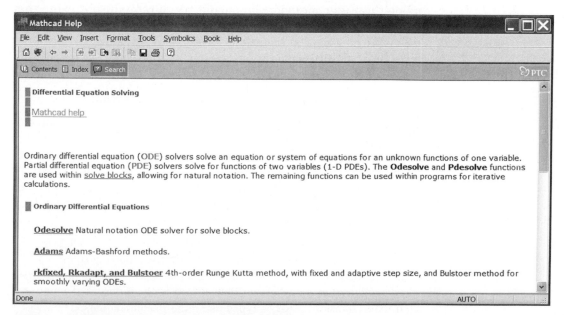

Figure 6.12: More information on ODE solvers

The next level of difficulty in solving ordinary differential equations is solving a *set of coupled first-order ODEs*. The methods we will discuss are:

1. The Euler method.
2. The *Odesolve* method.
3. The *rkfixed* method.

We will examine these methods by using each of them to solve the same problem. We illustrate with an example.

Example 6.6: Solving a Set of First-Order ODEs Solve the following set of equations over the range $t = 0$ to $t = 5$:

$$\frac{dx}{dt} = y(t)^2 - x(t) + e^{-t} \quad x(0) = 0$$

$$\frac{dy}{dt} = -x(t)z(t) \qquad y(0) = 2$$

$$\frac{dz}{dt} = y(t) - z(t) \qquad z(0) = 1$$

The Euler Method for a Set of First-Order ODEs

This method has already been discussed in some detail in Example 4.4. *You are urged to review that example before attempting to solve Example 6.6!* The worksheet shown in Fig. 6.13 shows the solution using the Euler method; you should try to reproduce it. Note that *all subscripts except* $_{end}$ *are array subscripts*, and that we *must* set up the three iteration equations for $x, y,$ and z in a matrix format to force their simultaneous computation, as illustrated in Fig. 6.14!

The Odesolve Method for a Set of First-Order ODEs

In Fig. 6.15 we show how to set up a worksheet using Odesolve to solve a set of first-order equations (Example 6.6). Comparing this worksheet to the one showing how to use Odesolve for solving one ODE, in Fig. 6.5, we see that solving a set of equations is almost as easy! You should be familiar with the worksheet of Fig. 6.5 (see Example 6.3) before attempting that in Fig. 6.15. We can make a few comments:

1. For the Odesolve you *must* start with the math region *Given*.
2. The ODEs to be solved *must* be typed with the following rules:
   a. The dependent variables must all be typed *with their independent variable arguments* wherever they appear in the differential equations.
   b. The equations *must* be typed using the Boolean equals.

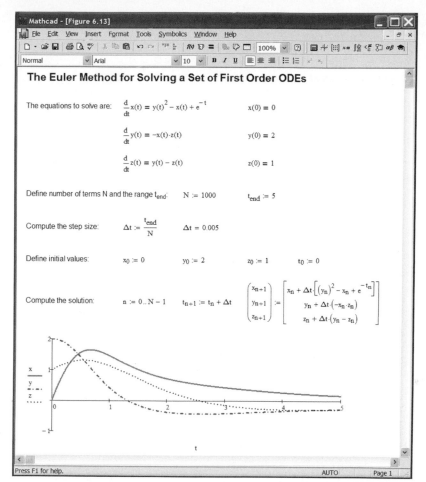

The Euler Method for Solving a Set of First Order ODEs

The equations to solve are:
$$\frac{d}{dt}x(t) = y(t)^2 - x(t) + e^{-t} \qquad x(0) = 0$$

$$\frac{d}{dt}y(t) = -x(t)\cdot z(t) \qquad y(0) = 2$$

$$\frac{d}{dt}z(t) = y(t) - z(t) \qquad z(0) = 1$$

Define number of terms N and the range t_{end}: $\quad N := 1000 \qquad t_{end} := 5$

Compute the step size: $\quad \Delta t := \dfrac{t_{end}}{N} \qquad \Delta t = 0.005$

Define initial values: $\quad x_0 := 0 \qquad y_0 := 2 \qquad z_0 := 1 \qquad t_0 := 0$

Compute the solution: $\quad n := 0..N-1 \qquad t_{n+1} := t_n + \Delta t$

$$\begin{pmatrix} x_{n+1} \\ y_{n+1} \\ z_{n+1} \end{pmatrix} := \begin{bmatrix} x_n + \Delta t \cdot \left[(y_n)^2 - x_n + e^{-t_n} \right] \\ y_n + \Delta t \cdot (-x_n \cdot z_n) \\ z_n + \Delta t \cdot (y_n - z_n) \end{bmatrix}$$

Figure 6.13:
The Euler method applied to a set of first-order ODEs

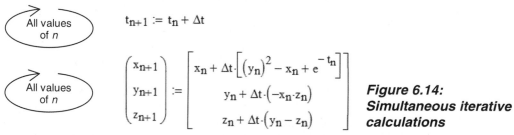

All values of *n*

$$t_{n+1} := t_n + \Delta t$$

All values of *n*

$$\begin{pmatrix} x_{n+1} \\ y_{n+1} \\ z_{n+1} \end{pmatrix} := \begin{bmatrix} x_n + \Delta t \cdot \left[(y_n)^2 - x_n + e^{-t_n} \right] \\ y_n + \Delta t \cdot (-x_n \cdot z_n) \\ z_n + \Delta t \cdot (y_n - z_n) \end{bmatrix}$$

Figure 6.14:
Simultaneous iterative calculations

c. The differential *must* be created using the ⧌ᵈ⁄ₓ icon in the *Calculus Toolbar* (or type **Shift + /**). However, instead of the differential, you may use the more compact notation involving the prime symbol, shown previously in Fig. 6.6. To get this, type ` (on the tilde key, ~), or type **Ctrl + F7**.

d. The initial conditions *must* be typed using the Boolean equals, and there must obviously be enough initial conditions to be able to solve the problem!

e. The Odesolve statement is used to define the solution as shown. The three arguments that *must* be used are: a column vector containing

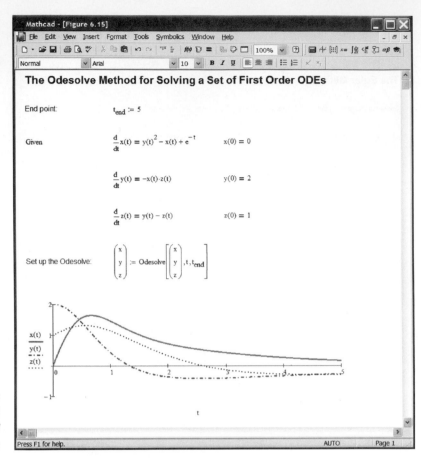

**Figure 6.15:
The Odesolve method
applied to a set of
first-order ODEs**

all the dependent variables (*without* their arguments); the independent variable (in this case t) and the end point (in this case t_{end}); an optional final argument (not used here) is the number of steps used in the solution (the default is 1000). Note that we set the result to be a column vector containing x, y, and z, *not* $x(t)$, $y(t)$, and $z(t)$, even though the statement *will* generate new functions, $x(t)$, $y(t)$, and $z(t)$! Actually, we could call the result any variable names; for example, we could use X, Y, and Z to generate solutions $X(t)$, $Y(t)$, and $Z(t)$.

3. As previously discussed, the Odesolve function uses a fixed step RK4 method to solve the differential equations, but you can right-click on the *Odesolve* and select *Adaptive* to make Mathcad automatically use smaller step sizes in regions where the functions are changing rapidly.

The rkfixed Method for a Set of First-Order ODEs

In Fig. 6.16 we show how to set up a worksheet using *rkfixed* to solve a set of first-order equations (Example 6.6). Comparing this worksheet to the one showing how to use *rkfixed* for solving one ODE, in Fig. 6.11, we see that solving a set of equations involves some special tricks, which we discuss below. You should be familiar with the worksheet of Fig. 6.11 (see Example 6.5) before attempting that in Fig. 6.16. We can make a few comments:

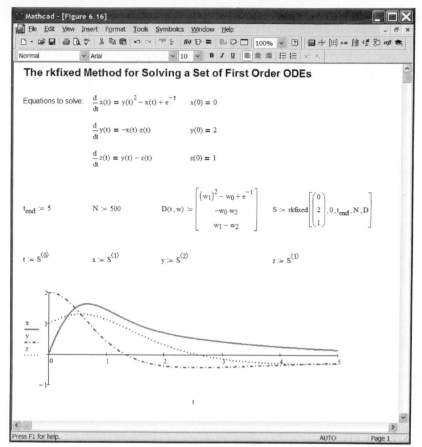

Figure 6.16:
The rkfixed method
applied to a set of
first-order ODEs

1. The differential equations and initial conditions are just for display purposes; they are not used.

2. All subscripts, except $_{end}$, are array subscripts—we are working with arrays here!

3. The *rkfixed* method will solve the ODEs at discrete points; in other words, it will not produce a continuous function, but instead a matrix of y values, for the solution. In the worksheet $N = 500$ is the number of points.

4. You *must* create a function $D(t,w)$ that is equal to the right-hand side of the standard ODEs, that is, the right-hand side of the ODEs after they are all written in standard form, $\frac{d}{dt} = \ldots$. This D function can be a bit tricky! The rules are:

   a. The function has two arguments: the independent variable (here t) and the dependent variable (here called w, but you can use any name). What is this w? It will represent x, y, and z!

   b. The right-hand side of the equation for D *must* be a column vector, with each element being one of the ODE right-hand sides, with the following special twist: *each variable to be solved for must be represented as one of the components of w!* In Fig. 6.16, comparing the three ODEs with the right-hand side of D, we see this means that x is represented by w_0, y is represented by w_1, and z is represented by w_2 (all array subscripts); and the equation for x is first, that for y is second, and that for z is last. Until you get the hang of it, it is a bit tricky!

5. You *must* create a new variable (in our worksheet S for "solution") and set it equal to the *rkfixed* function. The function is used in the following way:

***rkfixed*(initial values of w as a column vector, initial t, final t, number of points, D)**

Note that D must be entered without its arguments (i.e., D not $D(t,w)$).

6. The variable S is the solution to the set of differential equations! It is a matrix of values, *not* a function!

a. The first column of the matrix is the 501 values of the independent variable, in our case t (t_0 through t_N), generated by *rkfixed*.

b. The second column is the corresponding 501 values of the first dependent variable, in our case x (x_0 through x_N).

c. The third column is the second dependent variable, in our case y.

d. The fourth column is the third dependent variable, in our case z, which is also generated by *rkfixed* (you can verify all this by typing **S:** at the end of the worksheet).

7. To extract the values of the dependent and independent variables from S, we need to use the Matrix Column icon (M) on the Vector and Matrix Toolbar, or type **Ctrl + 6**. Note: In Fig. 6.16 the default value of *ORIGIN* is zero, so the first column of S is $<0>$; if you have *ORIGIN* set to unity, the first column will be $<1>$!

6.3

Numerically Solving a Higher Order ODE

Next we wish to see how to solve an ordinary differential equation of higher order. We will show how to solve a second-order equation—third-order, fourth-order, and so on equations (which actually don't really come up much in engineering science) can be solved using the same approach. We will consider:

1. The *rkfixed* method (and we'll hint at the Euler method).
2. The *Odesolve* method.

We will demonstrate these methods by analyzing Example 6.7.

Example 6.7: Solving a Higher Order ODE—Forced, Damped Vibration A mass $M = 1$ kg rests on an oiled surface (the friction coefficient is $c = 0.15$ N/(m/s)). The mass is attached to a wall using a spring, stiffness $k = 0.5$ N/m. A transient force $F(t)$, as shown in Fig. 6.17, is now applied. Find the position $x(t)$ of the mass over a 60-s period. The equation of motion for the mass is

$$\frac{d^2x}{dt^2} = \frac{1}{M}\left(F(t) - kx - c\frac{dx}{dt}\right)$$

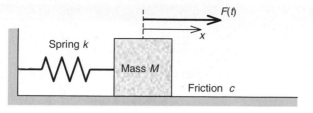

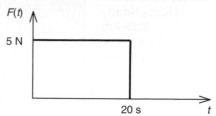

Figure 6.17:
Forced damped
vibration

The rkfixed Method for a Higher Order ODE

The *rkfixed* method is designed to solve one or more first order ODEs. How can we use it to solve a higher order ODE? Simple! We convert the higher order ODE into one or more first-order equations! For example, a second-order equation with initial conditions

$$\frac{d^2x}{dt^2} = f\left(t, x, \frac{dx}{dt}\right) \qquad x(0) = 0 \qquad \frac{dx}{dt}\bigg|_{t=0} = 0$$

can be written as two first-order equations

$$\frac{dx}{dt} = v(t) \qquad x(0) = 0$$

$$\frac{dv}{dt} = f(t, x, v) \qquad v(0) = 0$$

where we define a new dependent variable v to be the derivative of our original dependent variable, x. We have converted one second-order equation in one unknown to two first-order equations with two unknowns. (Similarly, a third-order equation could be rewritten as three first-order equations.)

The reason we do this is because we can now use *rkfixed* to solve these two first-order equations! For Example 6.7 we end up with

$$\frac{dx}{dt} = v \qquad\qquad\qquad x(0) = 0$$

$$\frac{dv}{dt} = \frac{1}{M}(F(t) - kx - cv) \qquad v(0) = 0$$

with the added insight that $v(t)$ is clearly not just an additional unknown but is actually the mass's velocity! We now have a straightforward set of first-order ODEs to solve (as in Example 6.6). You should be able to generate the solution worksheet shown in Fig.6.18; if you need help, see the discussion on the worksheet of Fig. 6.16, which also solves a set of first-order ODEs.

We won't do it here, but this logic (of breaking a higher order ODE into an equivalent set of first-order ODEs) can also easily be applied to the Euler method (we will use the Euler method for Example 6.7 in an exercise).

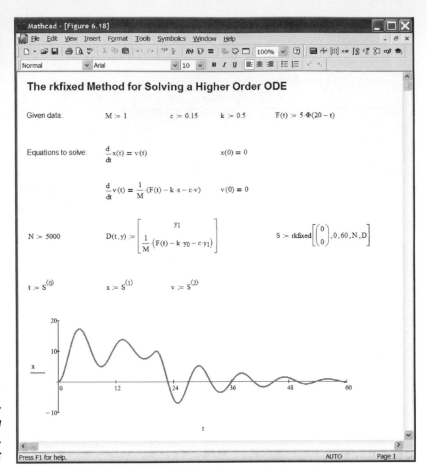

*Figure 6.18:
The rkfixed method
applied to a second-
order ODE*

The Odesolve Method for a Higher Order ODE

The Odesolve method is designed to solve first-order ODEs, or higher order ODEs! All we need to do is type the higher order equation after the Given, with sufficient initial conditions; we *don't* need to break the ODE down into a set of equivalent first-order ODEs! You should try to reproduce the worksheet shown in Fig. 6.19. If you need help, refer to the discussion of Example 6.4 (the worksheet shown in Fig. 6.9) for guidance.

Exercises

Note: Be aware that you may encounter problems if you solve a problem that involves a variable (say, n), and then use the same variable in a later problem in the same worksheet. If you get some strange results, this may be the reason; if so, make sure you reset the variable to a null value (e.g., set $n = 0$), or simply solve those problems on separate worksheets.

6.1 Consider a resistor-inductor (RL) circuit. The resistance is $R = 20$ ohm, and the inductance is $L = 6$ henry. The voltage applied to it is shown in

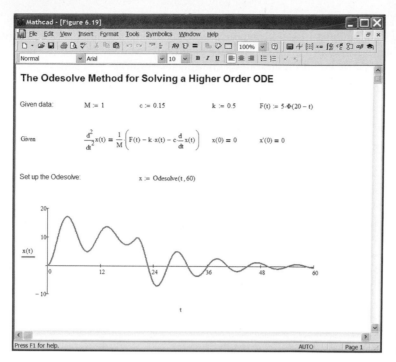

Figure 6.19:
The Odesolve method applied to a second-order ODE

Fig. E6.1. Find the current in the circuit as a function of time. The equation for this is

$$L\frac{dI}{dt} + RI = V(t)$$

Use the Euler and improved Euler methods.

6.2 The blood sugar level S in a patient is given by

$$\frac{dS}{dt} = I(t) - kS$$

where $I(t)$ is the sugar intake over 24 hours and $k = 0.15$ is a measure of the patient's natural sugar secretion. If the sugar intake over 24 hours is given by Fig. E6.2, plot the sugar level over the same period. The initial sugar level is 5.9 units. Use the Euler and improved Euler methods.

6.3 A robot cat is designed to chase a mouse, keeping away from the mouse a constant distance of $a = 10$ m. The cat runs directly toward the mouse at each instant. The initial positions of the cat and mouse are as shown in Fig. E6.3. If the mouse moves to the right along the x-axis, find and plot the path $y(x)$ that the cat takes. The equation for this is

$$\frac{dy}{dx} = -\frac{y}{\sqrt{a^2 - y^2}}$$

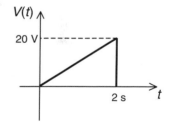

Figure E6.1

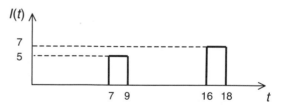

Figure E6.2

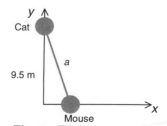

Figure E6.3

Use the Euler and improved Euler methods.

6.4 You are sitting in your car, which is in neutral. You change to drive and gradually press the accelerator pedal until you've "floored" it. It turns out that the equation describing the speed v (mph) is

$$\frac{dy}{dt} = F(t) - k_1 v - k_2 v^2$$

where $k_1 = 0.0075$ sec^{-1} and $k_2 = 0.0015$ 1/mph-sec are drag constants arising due to the rolling resistance of the tires and aerodynamic drag, and $F(t)$ is the thrust (mph/sec) produced by the engine, given by Fig. E6.4.

How long does it take you (within a tenth of a second) to reach 60 mph? What is your top speed (mph)? Use the Euler and improved Euler methods.

Hint: Note that the equation is written in such a way that you can enter the numerical values as given, and your answers will automatically be in mph. An equation for $F(t)$ can be written, using the Heaviside step function, based on Fig. E6.4. To find the time to reach 60 mph, the best approach is to plot the speed for the first 30 sec and then use the Trace feature. You can use Mathcad's max function to find the top speed.

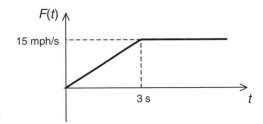

Figure E6.4

6.5 You are driving the car described in Exercise 6.4 at a steady speed of 60 mph (which requires a constant engine thrust $F_{engine} = 5.85$ mph/sec). After $T = 2$ seconds, you apply the brakes for a total time $\Delta T = 5$ sec Find the speed (mph) you slow down to while braking and how long it takes (within a tenth of a second) to speed up again to 55 mph after releasing the brakes. It can be shown that the equation for the car motion is

$$\frac{dv}{dt} = F_{engine} - k_1 v - k_2 v^2 - F_{brake} \; \Phi(t - T)\Phi(T + \Delta T - t)$$

where k_1 and k_2 are as before, and $F_{brake} = 10$ mph/sec is a factor representing the effect of applying the brakes (note that the product of Heaviside step functions has the effect of applying the brakes for the required period of time). Suppose you had *really* slammed on the brakes so that $F_{brake} = 15$ mph/sec. What would be your lowest speed and the time to speed back up to 55 mph? Use the Euler and improved Euler methods.

Hint: Review the hint given in Exercise 6.4. You can use Mathcad's min function to find the lowest speed. Don't forget to subtract $T + \Delta T$ from your answer for the time to pick up speed again!

6.6 Solve

$$y' = \sin(x) + y$$

with initial condition $y(0) = -\frac{1}{2}$, from $x = 0$ to 20. Compare this to the exact solution

$$y_{exact}(x) = -\frac{1}{2}(\sin(x) + \cos(x))$$

by plotting both on the same graph. Use the Euler and improved Euler methods.

6.7 Solve

$$y' = \frac{1}{4}(x + y)^2$$

with initial condition $y(0) = 0$, from $x = 0$ to $\pi/2$. Compare this to the exact solution

$$y_{exact}(x) = 2\tan\left(\frac{x}{2}\right) - x$$

by plotting both on the same graph. Use the Euler and improved Euler methods.

6.8 Solve

$$y' = y - y^2$$

with initial condition $y(0) = 2$, from $x = 0$ to 5. Compare this to the exact solution

$$y_{exact}(x) = \frac{2e^x}{2e^x - 1}$$

by plotting both on the same graph. Compute the maximum error (%). Use the Euler and improved Euler methods.

6.9 Solve

$$xy' = (x - y)^3 + y$$

with initial condition $y(1) = 1.5$, from $= 1$ to 2.235. Compare this to the exact solution

$$y_{exact}(x) = \frac{x}{\sqrt{5 - x^2}} + x$$

by plotting both on the same graph. Compute the maximum error (%). Use the Euler and improved Euler methods.

6.10 Solve the set of equations shown below

$$\frac{dx}{dt} = y$$

$$\frac{dy}{dt} = -|y|y - 5x$$

The initial conditions are $x(0) = 0$, $y(0) = 1$. Find the values of y and x at $t = 10$. Use the *rkfixed* and Odesolve methods (if necessary, increase the number of points, and use *Adaptive* elements).

6.11 Consider the circuit shown in Fig. E6.11, consisting of two inductors $L_1 = 5$ henry and $L_2 = 15$ henry, and two resistors $R_1 = 10$ ohm and $R_2 = 15$ ohm, initially with no currents flowing. At time $t = 0$ sec, a power surge given by

$$V(t) = V_{max} \sin\left(\frac{2\pi t}{T}\right)\Phi(T - t)$$

where $V_{max} = 120$ volt and $T = 1$ sec is applied. It can be shown that by summing the change in potential in each loop the following equations hold

$$V(t) = L_1\frac{dI_1}{dt} + R_1(I_1 - I_2)$$

$$0 = L_2\frac{dI_2}{dt} + R_2I_2 - R_1(I_1 - I_2)$$

Plot $I_1(t)$ and $I_2(t)$ (on the same graph) for the first 5 sec. Find the maximum currents flowing in each loop. Use the *rkfixed* and *Odesolve* methods (if necessary, increase the number of points, and use *Adaptive* elements).

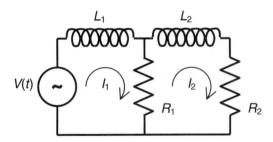

Figure E6.11

6.12 Consider the serial chemical reaction of three chemical species A, B, and C

$$A \xrightarrow{k_1} B \xrightarrow{k_2} C$$

where $k_1 = 0.2$ hr^{-1} and $k_2 = 0.1$ hr^{-1} are the reaction rate constants. It can be shown that the equations for the concentrations (mole/liter) of the species are

$$\frac{dA}{dt} = -k_1A$$

$$\frac{dB}{dt} = k_1A - k_2B$$

$$\frac{dC}{dt} = k_2B$$

If the initial concentrations are $A_0 = 15$ mole/liter, $B_0 = 10$ mole/liter, and $C_0 = 0$ mole/liter, plot a graph showing all three concentrations over the first 60 hours. Find the maximum concentration (mole/liter) of species B and when this occurs (hr). Use the *rkfixed* and *Odesolve* methods (if necessary, increase the number of points, and use *Adaptive* elements).

Hint: After solving the differential equations, you can use Mathcad's max function to find the maximum of B. One way to find the time is to use the Trace feature (assuming you computed a large enough number of data points to be reasonably accurate).

6.13 Repeat Example 6.7, but use the force shown in Fig. E6.13 instead. Solve using the *rkfixed* and *Rkadapt* methods and compare the results.

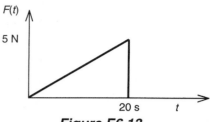

Figure E6.13

6.14 The pendulum can be made to swing by giving it an initial velocity. Suppose the instantaneous angle of the pendulum with the vertical is θ. With a high enough initial velocity, the pendulum can even be made to *rotate* through one or more complete revolutions, before friction slows it down. For $\theta(0) = 0$, find the integer initial velocities $\theta'(0)$ (e.g., $\theta'(0) = 1, 2$, etc.) at which the pendulum first swings completely over once, twice, three, and four times. To do this, solve using *Odesolve*, and plot $\theta(t)/(2\pi)$ (This will then be the angle θ in rotations rather than radians!) against time. The equation describing the motion is

$$\frac{d^2\theta}{dt^2} = -k\sin(\theta) - c\frac{d\theta}{dt}$$

where $k = 1$ depends on gravity and pendulum length, and $c = 0.25$ is a measure of air drag on the pendulum. Solve using the *rkfixed* and *Rkadapt* methods and compare the results.

6.15 Some physical systems have the property that for small oscillations energy is fed into them, but for larger amplitudes energy is removed. Such systems tend to seek a periodic behavior, although it will not be in any way sinusoidal. These systems can be modeled using the van der Pol equation

$$\frac{d^2x}{dt^2} - \mu(1 - x^2)\frac{dx}{dt} + x = 0$$

where x is the position at time t and $\mu = 2$. Find the behavior of the system over the time range $t = 0$ to 10 if it starts from rest at position $x = 2$. Plot x versus t, $v\left(= \frac{dx}{dt}\right)$ versus t, and v versus x, and discuss after searching the Internet for information on the van der Pol equation. Solve using the *rkfixed* and *Rkadapt* methods and compare the results.

6.16 Solve

$$y'' = xy' - 5y$$

with initial conditions $y(0) = 0$ and $y'(0) = 1$, from $x = 0$ to 9. Compare this to the exact solution

$$y_{\text{exact}}(x) = x - \frac{2}{3}x^3 + \frac{2}{30}x^5$$

by plotting both on the same graph. Compute the maximum error (%). (If you compare these results to those of Exercise 6.17, you can see how a minor change in a coefficient can have a radical effect on the answer.) Solve using the *rkfixed* and *Rkadapt* methods and compare the results.

6.17 Solve

$$y'' = xy' - 6y$$

with initial conditions $y(0) = 1$ and $y'(0) = 0$, from $x = 0$ to 9. Compare this to the exact solution

$$y_{exact}(x) = 1 - 3x^2 + x^4 - \frac{x^6}{15}$$

by plotting both on the same graph. Compute the maximum error (%). (If you compare these results to those of Exercise 6.16, you can see how a minor change in a coefficient can have a radical effect on the answer.) Solve using the *rkfixed* and *Rkadapt* methods and compare the results.

6.18 Solve

$$(1 - x^2)y'' - 2xy' + 6y = 0$$

with initial conditions $y(0) = -0.5$ and $y'(0) = 0$, from $x = 0$ to 20. Compare this to the exact solution

$$y_{exact}(x) = \frac{1}{2}(3x^2 - 1)$$

by plotting both on the same graph. Compute the maximum error (%). (The solution is the Legendre polynomial of degree 2.) Solve using the *rkfixed* and *Rkadapt* methods and compare the results.

6.19 Solve

$$(1 - x^2)y'' - 2xy' + 12y = 0$$

with initial conditions $y(0) = 0$ and $y'(0) = -1.5$, from $x = 0$ to 10. Compare this to the exact solution

$$y_{exact}(x) = \frac{1}{2}(5x^3 - 3x)$$

by plotting both on the same graph. Compute the maximum error (%). (The solution is the Legendre polynomial of degree 3.) Solve using the *rkfixed* and *Rkadapt* methods and compare the results.

6.20 Repeat Exercise 6.13 using the *Odesolve* method (if necessary, increase the number of points, and use *Adaptive* elements).

6.21 Repeat Exercise 6.14 using the *Odesolve* method (if necessary, increase the number of points, and use *Adaptive* elements).

6.22 Repeat Exercise 6.15 using the *Odesolve* method (if necessary, increase the number of points, and use *Adaptive* elements).

6.23 Repeat Exercise 6.16 using the *Odesolve* method (if necessary, increase the number of points, and use *Adaptive* elements).

6.24 Repeat Exercise 6.17 using the *Odesolve* method (if necessary, increase the number of points, and use *Adaptive* elements).

6.25 Repeat Exercise 6.18 using the *Odesolve* method (if necessary, increase the number of points, and use *Adaptive* elements).

6.26 Repeat Exercise 6.19 using the *Odesolve* method (if necessary, increase the number of points, and use *Adaptive* elements).

Importing, Exporting, and Analyzing Data

As with all Windows applications, Mathcad has numerous ways of communicating data between itself and other applications, and also with data files. In this chapter we will focus on exchanging data with two of the most common applications: Microsoft Word and Microsoft Excel. Other applications (such as MATLAB) follow similar procedures to one or the other of these two applications, so by considering just Word and Excel we will simplify the presentation without missing any important techniques. Once we learn how to communicate or transfer data back and forth from Mathcad, we will introduce some basic techniques for analyzing data, such as statistical analysis and trend line analysis. By "data" we usually mean numerical data, but in this chapter we also sometimes use it to refer to text, a table, a drawing in Word, or one or more math or text regions in Mathcad. Also note that we will use the following two techniques:

1. *Embedding*: We embed something when we paste it into the receiving application in such a way that it is still editable using the original application. *There is no permanent connection between the original file and the material embedded in the receiving application.* For example, if we embed a table from Word into Mathcad, we will have within the Mathcad worksheet a region that looks just like the original Word data and that can be edited by double-clicking (which temporarily replaces Mathcad's icons with Word's). Any changes to the embedded Word data in the Mathcad worksheet do *not* show up in the original Word file, and changes to the original Word file likewise do *not* affect the embedded data.

2. *Linking*: We link something when we paste it into the receiving application in such a way that the data are still connected to the original application. *There is a permanent connection between the original file and the material linked in the receiving application.* For example, if we link a table from Word into Mathcad, we will have within the Mathcad worksheet a region that looks just like the original Word data. It is edited by double-clicking, which transfers the user to the *original* Word document for editing.

Transferring Data between Mathcad and Word

To transfer data from Word to Mathcad or from Mathcad to Word, we have the following options available:

1. Drag and drop (utilizing OLE 2).

2. Copy and paste, with or without embedding. As discussed above, embedding creates a new Word document that contains the data *within* the Mathcad worksheet and leaves the original Word document intact and unconnected to the Mathcad worksheet.

3. Copy and paste special, with or without linking. As we also discussed above, linking creates a Word region containing the data *within* the Mathcad worksheet; *any future changes to the original Word document appear in the Mathcad worksheet.*

To be honest, Mathcad now has its own excellent formatting features; there is usually not much of a need to feed data (e.g., text) from Word to Mathcad, with the exception of drawings done using Word's Drawing Toolbar. On the other hand, going from Mathcad to Word is a useful thing to do because Word's computation capabilities (yes, Word has them for use within tables) are limited. We illustrate the methods with an example.

Example 7.1: The Mass of Various Cylinders A Word document contains a drawing of and data (in mixed SI and English units) on several cylinders. Find the mass (lb) of each cylinder and present the results in the Word document.

The original Word document and a blank Mathcad worksheet are shown in Fig. 7.1. Note that Fig. 7.1 shows Word's Drawing Toolbar in the lower left—you are urged to become familiar with it because it's an excellent little tool for drawing basic figures! You might want to practice its use (including use of the 3D Style icon) by trying to reproduce the drawing of the cylinder. For this example, you should also try to reproduce the rest of the Word document, and arrange Word and Mathcad windows as shown in the figure. The first method we will consider is drag and drop.

Drag and Drop (Embedding Using OLE 2)

We can drag and drop any, or even all, of the content of a Word document, simply by selecting it in the Word window and dragging into the Mathcad window: the content will be *embedded*. There is one catch: the content will disappear from the Word document! To make the content reappear in Word (but still keep the copy in Mathcad), you can click on the Undo Typing icon on Word's

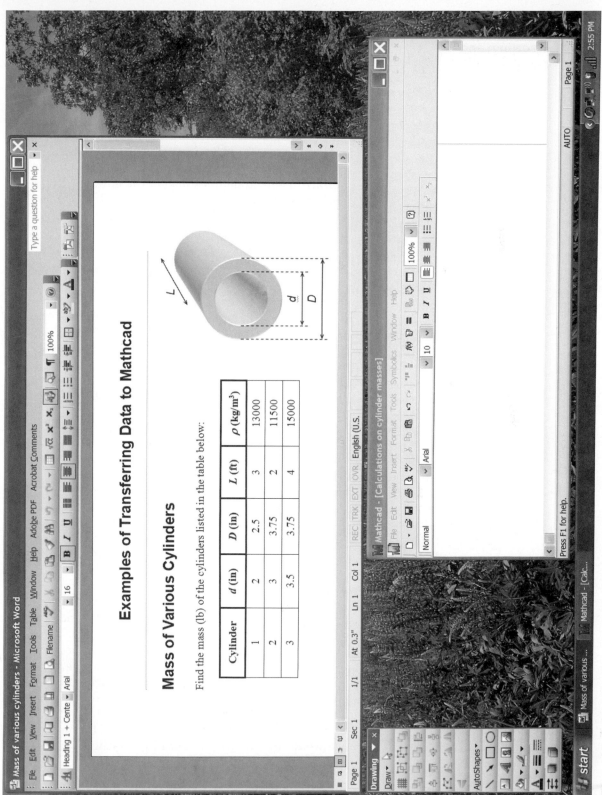

Figure 7.1: Transferring data using drag and drop (OLE 2)

Standard Toolbar; better yet, *to avoid losing the original Word content in the first place*, while dragging and dropping hold down the **Ctrl** key. As explained above, embedding creates a new Word document containing the data *within* the Mathcad worksheet, *and the original Word document is unconnected to the Mathcad worksheet*.

For example, if you are in the Word window, you can select, say, part of the table by wiping over it; then hold the **Ctrl** key while dragging your selection into the Mathcad window. You will end up with a Word region in the Mathcad worksheet that contains the selected material. To edit it, double-click on the data to get the Word interface within Mathcad! You can even type **Ctrl + A** within the Word document to select the *entire* document and then drag it into the Mathcad worksheet; it will appear as a single Word region within Mathcad, which can be edited by double-clicking.

Suppose in this exercise we wish to have a copy of the drawing in the Mathcad worksheet in which we will do the calculations. To create a copy in Mathcad, we simply select the Word drawing by clicking on the edge of it, and then drag the drawing from the Word window to the Mathcad window. There is one catch (which in my view is really a bug in Word and/or Mathcad): the drawing generated in Mathcad is a graphic that cannot be edited and is no longer connected to Word in any way. If you want an *editable* drawing in Mathcad, you need to use a little trick: drag and drop the drawing *and* a piece of adjacent text (any text, even a single letter). After inserting the Word region in Mathcad you can then double-click it to enter Word edit mode and delete the unwanted text! As an exercise, you can try transferring a copy of your drawing to the Mathcad worksheet.

You can also use this method to transfer data from Mathcad to Word: simply select the Mathcad regions you wish to transfer, click with the mouse in one of the regions to get the small black hand (✋); then (holding the **Ctrl** key as desired) drag from Mathcad to the Word document. When you switch to the Word document, you'll have a Mathcad file embedded in the document. If you double-click on this, you'll be working on a *complete* copy of the original Mathcad worksheet that is embedded in the Word document. Because it's an embedded Mathcad region in Word, the original Mathcad worksheet is unaffected and not connected to the Word document.

This method of embedding data from Word to Mathcad or Mathcad to Word by simply dragging and dropping is very convenient when you have a small amount of data to transfer, but it has some limitations. For example, we would like to get data from the table in the Word document shown in Fig. 7.1 into the Mathcad worksheet in a way that Mathcad can be used to compute with the data; we would like to *use* the data by creating a matrix *M*. You can't do this with this drag and drop method; however, you can do it with the next method.

Copy and Paste, or Copy and Paste Special (Embedding or Linking)

Using copy and paste or copy and paste special is probably the easiest, most useful and predictable way for transferring data between Word and Mathcad. We can illustrate copying and pasting from Word to Mathcad using the Word table of Example 7.1. Select the three rows and four columns containing the data on cylinder diameters, lengths, and densities, and copy. Next, switch to Mathcad, in which you can use various options for pasting.

1. *Paste* directly onto the worksheet (by right-clicking on an empty region and selecting *Paste*, by typing **Ctrl + V**, or by using *Edit . . . Paste*). In this method, a Word region containing the selected data is *embedded* in Mathcad.

2. *Paste Special* directly onto the worksheet (by right-clicking on an empty region and selecting *Paste . . . Special* or by using *Edit . . . Paste Special*). In this method, we can make a series of choices, as shown in Fig. 7.2. Most of these are self-explanatory (and not that useful), but we point out two:

    a. *Unformatted Text as Math*: This option will paste the data as math; for example, the table data we selected will appear as a matrix.

    b. *Microsoft Office Word Document*: This option inserts the data as an embedded Word region within Mathcad; if you click on Paste Link first, the window changes to that shown in Fig. 7.3, and the data will be pasted as a linked Word region—double-clicking on the region allows you to edit the original Word document.

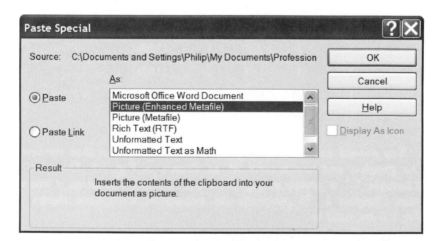

Figure 7.2: The Paste Special window for a Word paste

You may wish to verify the methods for pasting directly to Mathcad by practicing each of them on a blank worksheet.

These methods are okay, but they paste the data in such a way that Mathcad itself can't use the data! A more useful method is to paste into an equation *placeholder* rather than to an empty region in Mathcad. For example (assuming you copied just the

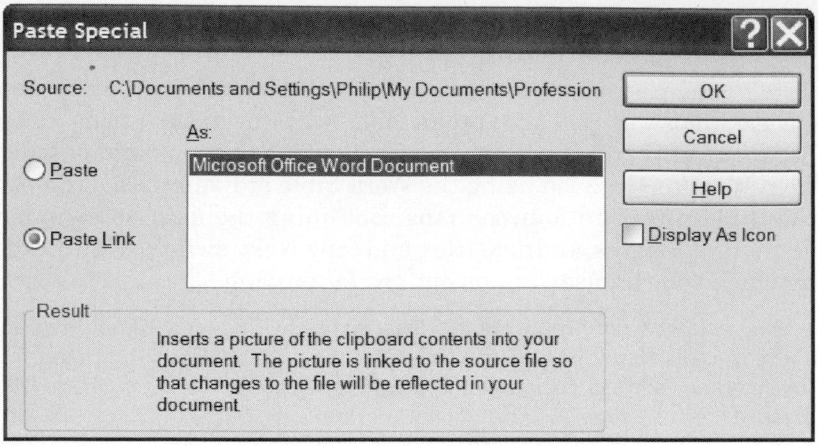

Importing, Exporting, and Analyzing Data

three rows and four columns containing the data on cylinder diameters, lengths, and densities), in Mathcad we can type **M:** and then immediately paste (e.g., by typing **Ctrl + V**); we get a useful matrix of the values. (We could also have selected the entire table in the Word document, and pasted this into the placeholder. This would produce a matrix as shown in Fig. 7.4; note that the matrix contains numerical values as well as strings!)

Figure 7.4:
An example of pasting a Word table into a Mathcad placeholder

$$M := \begin{pmatrix} \text{"Cylinder"} & \text{"d (in)"} & \text{"D (in)"} & \text{"L (ft)"} & \text{"}\square\text{ (kg/m3)"} \\ 1 & 2 & 2.5 & 3 & 13000 \\ 2 & 3 & 3.75 & 2 & 11500 \\ 3 & 3.5 & 3.75 & 4 & 15000 \end{pmatrix}$$

You should now be able to generate the worksheet that contains the graphic and the matrix M, as shown in Fig. 7.5. The rest of Example 7.1 is pretty straightforward (although it helps if you've read Chapter 4). To generate the completed worksheet shown in Fig.7.6 we need to:

1. Extract the values of cylinder diameters d and D, length L, and density ρ; we also need to assign to them dimensions of in (for d, D, and L) and kg/m^3 (for ρ). This is accomplished using the four equations shown in Fig. 7.6. The function submatrix(M, 0, 2, 2, 2), for example, creates a new matrix by extracting from M the submatrix defined by rows 0 through 2, and columns 2 through 2 (i.e., column 2 only). Note that ORIGIN is set at default value zero here! As we define each variable, we do so with the units we desire.

2. Evaluate diameters d and D, length L, and density ρ to make sure everything is correct.

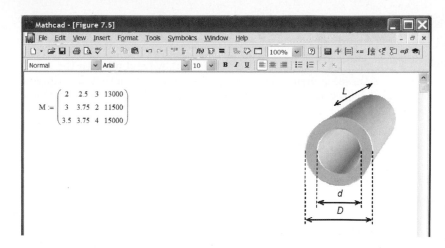

Figure 7.5:
The worksheet after
pasting a graphic
and some data into a
placeholder

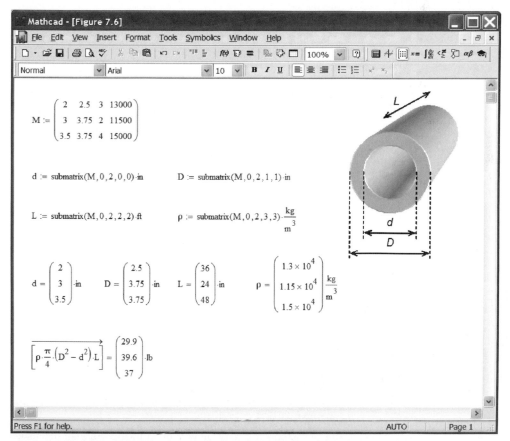

Figure 7.6: The completed worksheet for Example 7.1

3. Compute the mass of each cylinder. Note that we *vectorized* (by using the 🔃 icon on the Vector and Matrix Toolbar or by typing **Ctrl + -**) the equation for the cylinder masses, which causes the operation to be done on each element of the vectors! Figure 7.7 shows what would happen if we didn't vectorize: we would get an answer, but it would be incorrect! It evaluates $D^2 - d^2$ correctly, producing a new 3×1 matrix, but when we then premultiply by ρ (another 3×1 matrix), we end up with the scalar product of these—a meaningless computation in this context. Multiplying this scalar by L (yet another 3×1 matrix) generates the (incorrect) 3×1 answer. For our correctly vectorized result, we inserted units of lb, as desired.

**Figure 7.7:
Evaluating the masses
incorrectly**

$$\rho \cdot \frac{\pi}{4} \cdot \left(D^2 - d^2\right) \cdot L = \begin{pmatrix} 117 \\ 78 \\ 156 \end{pmatrix} \text{lb}$$

To complete Example 7.1, we need to get the results for the mass back into the original Word document. The easiest way to do this is to:

1. Select and copy the data, as shown in Fig. 7.8.

2. Switch to the Word document, and create a table with an empty 3×1 set of cells.

3. Select those Word cells, and paste.

The final Word document is shown in Fig. 7.9. Once again, Mathcad proved very useful for performing computations, especially with the mixed units!

**Figure 7.8:
Selecting data to be
copied**

$$\left[\rho \cdot \frac{\pi}{4} \cdot \left(D^2 - d^2\right) \cdot L\right] = \begin{pmatrix} 29.9 \\ 39.6 \\ 37 \end{pmatrix} \cdot \text{lb}$$

7.2

**Transferring Data
between Mathcad
and Excel**

To transfer data from Excel to Mathcad or from Mathcad to Excel we have even more options than we had with Word. In the next few pages, we first review the same options we had in Word.

1. Drag and drop (utilizing OLE 2).

2. Use copy and paste, with or without embedding. As discussed above, embedding creates a new Excel worksheet containing the data *within* the Mathcad worksheet and leaves the original Excel worksheet intact and unconnected to the Mathcad worksheet.

3. Use copy and paste special, with or without linking. As also discussed above, linking creates an Excel worksheet containing the data *within* the Mathcad worksheet; *any future changes to the original Excel worksheet appear in the Mathcad worksheet.*

Then we can look at some much better special methods that don't apply when working with Word. These are:

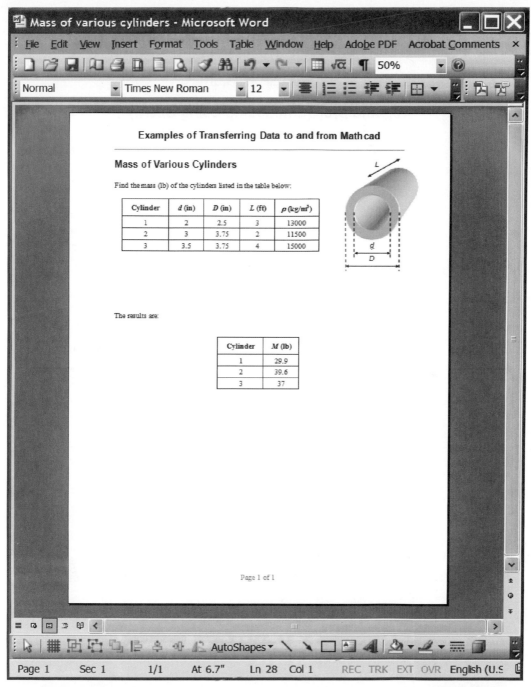

Figure 7.9:
The completed Word document

1. Inserting an Excel component. This is a very useful way to set up communication between Mathcad and Excel. Using it sets up a region in Mathcad that holds an *embedded* Excel worksheet, so it is best for use if you're working with both applications but your main focus is Mathcad.

137

2. Using the Mathcad add-in in Excel (in later Mathcad 14 versions). This sets up a region in Excel that holds an *embedded* Mathcad worksheet; it is best for use your if main focus is Excel.

First, we review the methods already discussed in detail when using Word for the same procedures.

Drag and Drop (Embedding Using OLE 2)

As we discussed in detail above with Word, you can drag and drop content of an Excel worksheet simply by selecting it in the Excel window and dragging into the Mathcad window. The content will be *embedded*—as we saw with Word. *To avoid losing the original Excel content*, hold down the **Ctrl** key while dragging and dropping. This embeds a new Excel worksheet containing *all* the data in the original Excel worksheet *within* the Mathcad worksheet; *the original Excel worksheet is unconnected to the Mathcad worksheet*.

Copy and Paste, or Copy and Paste Special (Embedding or Linking)

Once again, we have a method similar to what we discussed above in detail for Word. After copying data from an Excel worksheet:

1. *Paste* directly onto a Mathcad worksheet (by right-clicking on an empty region and selecting *Paste*; by typing **Ctrl + V**; or using by using *Edit . . . Paste*). In this method, an Excel region showing the selected data is *embedded* in Mathcad; actually, an *entire* copy of the Excel worksheet is embedded.

2. *Paste Special* directly onto the Mathcad worksheet (by right-clicking on an empty region and selecting *Paste . . . Special*; or by using *Edit . . . Paste Special*). In this method, we have a series of choices we can make, as shown in Fig. 7.10. Most of these are self-explanatory; the most useful are:

    a. The *Unformatted Text as Math* option (not visible in Fig. 7.10, but in the list) will paste the data as a Mathcad matrix.

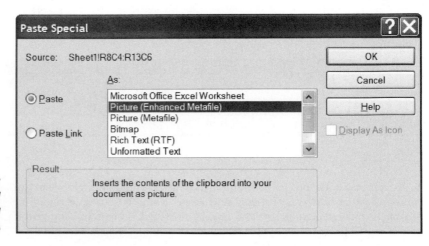

*Figure 7.10:
The Paste Special
window for an Excel
paste*

b. The *Microsoft Office Excel Worksheet* option inserts the data as an embedded Excel region within Mathcad. If you click on *Paste Link* first, the window changes to that shown in Fig. 7.11, and the data will be pasted as a linked Excel region—double-click on the region to edit the original Excel worksheet.

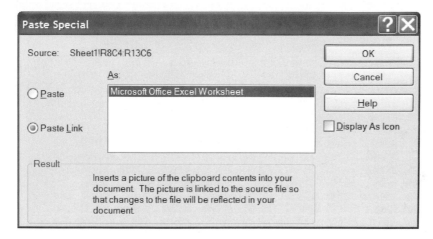

You may wish to verify these methods for pasting from Excel to Mathcad by practicing each of them using arbitrary data in an Excel worksheet, and transferring to a blank worksheet. *Note that copying data from Mathcad to Excel follows the same basic procedures as just described for copying data from Excel to Mathcad (except linking does not appear to be an option).*

These methods are okay, but they don't give you anything in Mathcad that you can *use* in Mathcad! A better method is to paste the Excel data into an equation *placeholder* rather than to an empty region in Mathcad. We will illustrate this with an Example, using the Excel worksheet shown in Fig. 7.12. It's a good idea to try to generate a similar worksheet so that you can follow along.

To generate the (fake) production data we used the following formula in cells C4 through I30: **=INT(NORMINV(RAND() ,750,50))**. This computes an integer random number based on a normal distribution whose mean is 750 and standard deviation is 50. To keep Excel from recomputing new values each time we do anything in Excel, we copied all the values (using **Ctrl + C**) and then immediately right-clicked cell D5 to get Excel's *Paste Special* . . . to paste just the values. (Of course, your particular values will be a different random collection.)

Example 7.2: Analysis of Production Data An Excel worksheet contains data on the output of a coffee machine manufacturer that has ten production plants. Using copy and paste between Excel and Mathcad, (a) find the maximum, minimum, mode, and mean of all the plants' production for the year; (b) plot a histogram of the production for January for all

File   Edit   View   Insert   Format   Tools   Data   Window   Help   Adobe PDF

Type a question for help

Arial   ▼ 10 ▼   B   I |≡ ≡ ≡ ≣| $ %   ⁺⁰ ⁰⁰ | ⊞ ▼ ◇ ▼ | ⊞ ⊕ ⊼ ✕ ⊟ | ◈ ✎ | [?]   Mathcad ▼

A1   ▼   fx

Production Data on Coffee Machine Factories

Month	Plant 1	Plant 2	Plant 3	Plant 4	Plant 5	Plant 6	Plant 7	Plant 8	Plant 9	Plant 10
Jan	844	754	777	701	749	799	732	811	782	742
Feb	669	800	655	704	732	682	714	718	762	714
Mar	784	814	794	816	779	786	774	694	687	823
Apr	789	783	693	807	763	814	739	728	760	780
May	732	751	725	733	733	713	849	733	809	756
Jun	774	802	697	667	801	522	722	840	747	759
Jul	699	800	716	759	752	707	753	715	863	773
Aug	811	751	740	754	814	722	775	757	783	699
Sep	750	730	680	754	720	848	773	731	798	659
Oct	752	724	775	665	742	761	753	714	662	801
Nov	770	770	723	807	693	809	708	859	776	704
Dec	714	748	693	769	779	725	704	706	761	726

Sheet1 / Sheet2 / Sheet3 /

Figure 7.12:
The Excel file for
Examples 7.2, 7.3,
and 7.4

plants' production; (c) plot a histogram for Plant 1 production for the year; and (d) plot a histogram of all the plants' production for the year.

All parts of this example can actually be done with Excel itself, but we use Mathcad so we can practice moving data between Excel and Mathcad, as well as learning a little about some Mathcad statistical functions.

Assuming you have a version of the Excel worksheet shown in Fig. 7.12, we are now ready to practice copying and pasting into a Mathcad placeholder. The procedure couldn't be much simpler.

1. Copy the cells in Excel containing the data.

2. Switch to Mathcad, type **P:** (or any other variable name you wish), and then immediately paste (e.g., by typing **Ctrl + V**); we get a useful matrix of all the values.

Figure 7.13 shows the first part of the solution to Example 7.2. The large matrix P contains all the values from the Excel worksheet!

The worksheet also shows some statistical and array calculations. For this example, we:

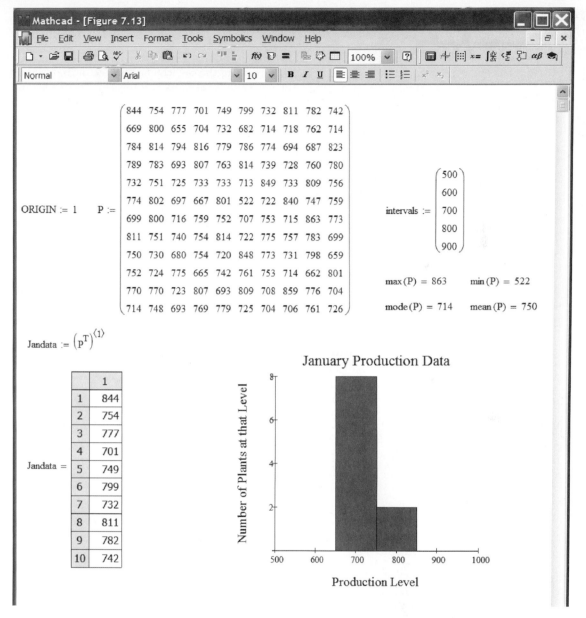

Figure 7.13:
The first part of the
Mathcad file for
Example 7.2

1. Set the array *ORIGIN* value to unity.

2. Use Mathcad functions to obtain the maximum, minimum, mode, and mean of *P*.

3. Define an array, *intervals*, which will be used for the histogram calculation. This is a column vector of a size of our choosing, with values of our choosing. In this case, we chose it to be a 5 × 1 column, with values as shown. These values will be used by Mathcad as the "bins" into which data will be collected (see point 5 below).

4. Extract the first row of data from P (which is the January data for all plants). We can't directly extract a row of data from a matrix in Mathcad (unless we use array subscripts), but we *can* extract a column (by using the M° icon on the Vector and Matrix Toolbar or by typing **Ctrl + 6**); hence, we take the *transpose* of P (by using the Mᵀ icon on the Vector and Matrix Toolbar or by typing **Ctrl + 1**) so that columns and rows are switched, and then we extract the first column!

5. Plot a histogram of the data. The information plotted in the graph is shown in Fig. 7.14. Mathcad has a function hist(*intvls*, *data*), which uses a column vector (*intvls*) to define the "bins" into which the values in *data* are counted. In our case, we use the variable *intervals* to define bins 500, 600, 700, 800, and 900, and instead of *data,* we define *Jandata*. With these arguments, hist counts how many of values of *Janadata* are in the 500s (zero, it turns out), 600s (zero again), 700s (eight) 900s (two), and 1000s (zero). Details on graphing are discussed in Appendix: Graphing, but for now:

a. Click *Insert . . . Graph . . . X-Y Plot* (or use the ⬔ icon on the *Graph Toolbar*) or type **@**.

b. On the middle and right placeholders on the horizontal axis, type **intervals** and **1000** (otherwise the graph will plot only to 900).

c. Click on middle black placeholder of the vertical axis and type **hist(intervals,Jandata)**.

d. Press **Enter** to exit the graph.

e. Double-click on the graph to make the axis style crossed, to format the trace type as *solidbar*, to hide the arguments and legend, and to enter the labels shown.

You should be able to finish the rest of Example 7.2; it looks something like Fig. 7.15. We can now explore some slightly more sophisticated methods for connecting Mathcad and Excel.

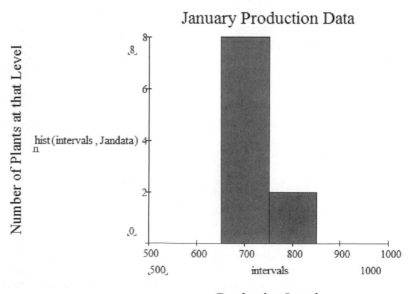

**Figure 7.14:
Formatting the graph
for Example 7.2**

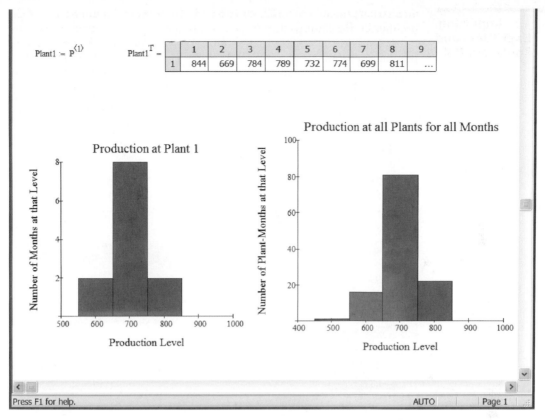

Plant1 := $P^{\langle 1 \rangle}$

Plant1^T =		1	2	3	4	5	6	7	8	9
	1	844	669	784	789	732	774	699	811	...

Figure 7.15:
The last part of the Mathcad file for Example 7.2

Inserting an Excel Component in Mathcad

This is a method specially designed for use in Mathcad, for getting data *into* Mathcad *from* Excel as well as *from* Excel *to* Mathcad. We italicized several words to indicate the flow of information, and also to focus on what I think is a slight design error in Mathcad. In using the Excel component, the designers of Mathcad decided to call *Inputs* data that is moving from Mathcad to Excel, and *Outputs* data that is moving from Excel to Mathcad. To me, this is kind of a backwards notation—if you have an object for transferring data in a Mathcad worksheet, shouldn't data *entering* Mathcad be called input, not output? The notation used in Mathcad is worth repeating:

1. *Inputs*: Data that is moving from Mathcad to Excel.
2. *Outputs*: Data that is moving from Excel to Mathcad.

We demonstrate use of an Excel component via an example.

Example 7.3 (Example 7.2 *Revisited*): Analysis of Production Data An Excel worksheet contains data on the output of a coffee machine manufacturer that has ten production plants. By inserting an Excel component in Mathcad, (a) find the maximum, minimum, mode and mean of all the plants' production for the year; (b) plot a histogram of the production for January for all plants' production; (c) plot a histogram for Plant 1 production for the year; and (d) plot a histogram of all the plants' production for the year.

To start this example we need an Excel file similar to that shown in Fig. 7.12. To generate your own version of the file, see the description in the paragraph immediately before Example 7.2.

First, we review the steps necessary for inserting an Excel component.

1. Click on menu item *Insert . . . Component*, to get the window shown in Fig. 7.16. Note that there are a number of options here! We discuss two of them (the others have a similar logic).

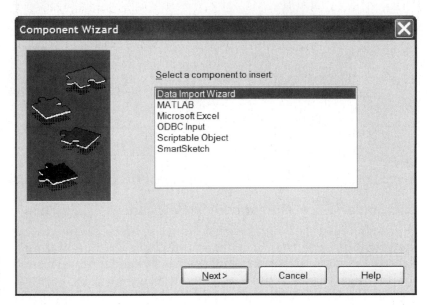

*Figure 7.16:
The Component
Wizard window*

a. The *Data Import Wizard* option leads to the window shown in Fig. 7.17. In this window you can select whether the data it imports (from a text file, e.g., *.dat, *.csv, *.txt) should be assumed to be separated by tabs, commas, or spaces, and so on. You can also decide if the component should be displayed in Mathcad as a table or as an icon. Using this option to import data from an existing text file is straightforward (We are using a "Wizard" after all!), so we won't explore it any further here.

b. The *Microsoft Excel* option leads to the Excel Setup Wizard window shown in Fig. 7.18. Note that once again you can choose to have the component displayed in Mathcad as a table or as an icon.

2. Select *Create from an empty Excel worksheet* to set up a new, blank worksheet to embed or *Create from file* to embed an existing file. (When we solve Example 7.3, we will choose the Excel file shown in Fig. 7.12.)

3. With either choice in step 2, we get a second Excel Setup Wizard window, similar to that shown in Fig. 7.19. Keeping in mind what we mean by *Inputs* and *Outputs*, we can specify:

Figure 7.17:
The Data Import
Wizard window

a. How many Mathcad variables we will be using to send data *to* Excel (*Inputs*). You don't *have* to have inputs—you can set the number of *Inputs* to zero. For each input you must specify the *top left corner cell* on the Excel worksheet where Mathcad will place the data. This obviously involves some planning if you're creating a new Excel worksheet, or knowledge of the layout of an existing worksheet, especially if you intend to send a lot of data, or a large number of variables, to Excel.

b. How many Mathcad variables we will be using to receive data *from* Excel (*Outputs*). Once again, you don't *have* to have outputs—you can set the number of *Outputs* to zero. You must specify the *complete range* from which Mathcad will retrieve the data for each variable (if you know

Figure 7.18:
The first Excel Setup
Wizard window

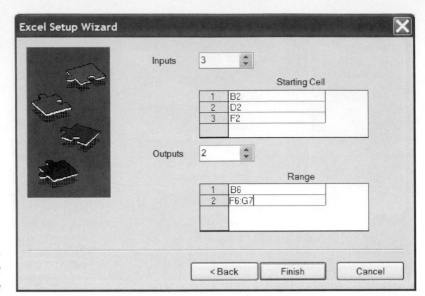

how to name ranges in Excel, you can just use the range name instead), and so you again need to know the layout of the Excel worksheet.

4. Decide whether to have the component appear within Mathcad as a table or as an icon.

5. Click *Finish* to return to the Mathcad worksheet. There will be blank placeholders for you to enter the names of the relevant *Input* and *Output* variables: *the placeholders at the top left are always the* Outputs (data from Excel to Mathcad), and *the placeholders at the bottom left are always the* Inputs (data from Mathcad to Excel).

This procedure (especially step 3) seems a bit complicated, but Figs. 7.19 and 7.20 demonstrate how it works. (You may want to try to reproduce Figs. 7.19 and 7.20 to practice the technique.) Figure 7.19 shows an example where three inputs were selected; each set of data from Mathcad was inserted into a blank Excel worksheet, starting at cells B2, D2, and F2. We created two outputs for sending data from Excel back to Mathcad; one from the single cell B6 and the second from the 2×2 range F6:G7. Figure 7.20 shows the Mathcad worksheet into which this Excel component was inserted. First, three variables x, y, and z were defined in Mathcad. Then the component was inserted, and it had five empty placeholders: two at the top left for *Outputs* (data from Excel to Mathcad) and three at the bottom for *Inputs* (data from Mathcad to Excel). In Fig. 7.20 we used the previously defined variables x, y, and z as outputs and defined new variables A and B to receive whatever we computed in the Excel worksheet in cell B6 and range F6:G7. Just as an exercise, we double-clicked on the component to access the embedded Excel worksheet, and then typed into cell B6 the formula **=SUM(D2:D4)** to add the three y values. In cell F6 we typed **=F2^2** (and then copied and pasted to range F6:G7) so that each cell contained the square of each cell in the range obtained

$$x := 1 \qquad y := \begin{pmatrix} 1 \\ 4 \\ -2 \end{pmatrix} \qquad z := \begin{pmatrix} 2 & 1 \\ 4 & 3 \end{pmatrix}$$

$$\begin{pmatrix} A \\ B \end{pmatrix} :=$$

	(This is x)		(This is y)		(This is z)	
	1		1		2	1
			4		4	3
			-2			
	(This is A)				(This is B)	
	3				4	1
					16	9

$(x \ y \ z)$

$$A = 3 \qquad B = \begin{pmatrix} 4 & 1 \\ 16 & 9 \end{pmatrix}$$

*Figure 7.20:
The Mathcad
calculations
corresponding to
Fig. 7.19*

from variable z. Finally, in the Mathcad worksheet, we evaluated the variables A and B. If you check the numbers shown, you can see that Figures 7.19 and 7.20 demonstrate that we can pass data (x, y, z) from Mathcad to Excel; perform calculations in Excel; and feed data (A, B) back to Mathcad. If we change any values of x, y, or z, the changes immediately get sent to Excel via the component, calculations in Excel are updated, and new values are sent to A and B in Mathcad. We repeat a final note here: remember that you don't have to have any *Output*, or any *Inputs* (but obviously you do need at least one of one of them).

Returning to the solution of Example 7.3, we could create an Excel component that has three outputs: P (for the entire table of data in Fig. 7.12), *Jandata* (for extracting just January data), and *Plant1* (for extracting just the data on plant 1). (An alternative would be to just have one output, P, and have Mathcad do the extracting of data as in Example 7.2.)

1. Insert an Excel component into a new Mathcad worksheet.

2. Select the *Create from file* option in the first setup window (Fig. 7.18).

3. Browse for the Excel file (Fig. 7.12).

4. Fill in the data in the second setup window as shown in Fig. 7.21, and click on *Finish*.

5. Enter the three variable names in the placeholders for outputs.

We end up with the first part of the solution to Exercise 7.3 shown in Fig. 7.22. The component is displayed in Mathcad, with its three outputs (and no inputs).

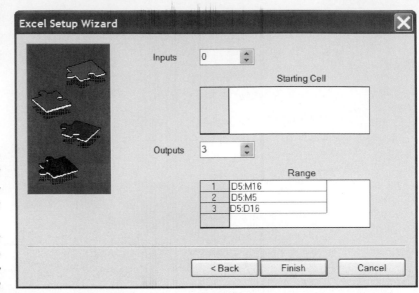

Figure 7.21:
The second Excel
Setup Wizard window
for Example 7.3

Figure 7.22:
The first part of the
Mathcad file for
Example 7.3

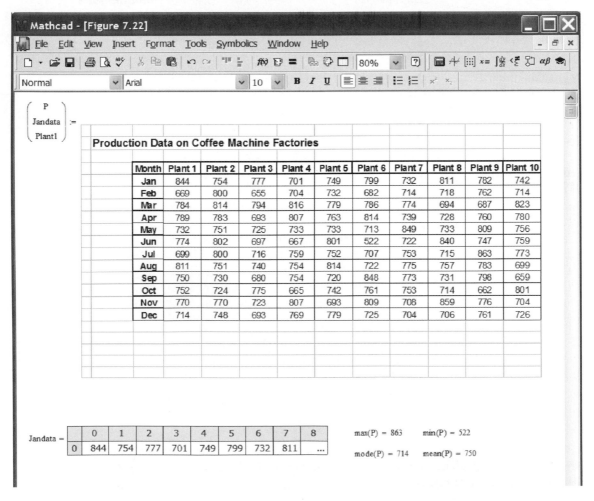

$$\begin{pmatrix} P \\ Jandata \\ Plant1 \end{pmatrix} :=$$

Production Data on Coffee Machine Factories

Month	Plant 1	Plant 2	Plant 3	Plant 4	Plant 5	Plant 6	Plant 7	Plant 8	Plant 9	Plant 10
Jan	844	754	777	701	749	799	732	811	782	742
Feb	669	800	655	704	732	682	714	718	762	714
Mar	784	814	794	816	779	786	774	694	687	823
Apr	789	783	693	807	763	814	739	728	760	780
May	732	751	725	733	733	713	849	733	809	756
Jun	774	802	697	667	801	522	722	840	747	759
Jul	699	800	716	759	752	707	753	715	863	773
Aug	811	751	740	754	814	722	775	757	783	699
Sep	750	730	680	754	720	848	773	731	798	659
Oct	752	724	775	665	742	761	753	714	662	801
Nov	770	770	723	807	693	809	708	859	776	704
Dec	714	748	693	769	779	725	704	706	761	726

Jandata =

	0	1	2	3	4	5	6	7	8
0	844	754	777	701	749	799	732	811	...

$\max(P) = 863$ $\min(P) = 522$

$\mathrm{mode}(P) = 714$ $\mathrm{mean}(P) = 750$

Once we get the data into Mathcad, the solution (which we leave to you) to Example 7.3 is almost identical to that for Example 7.2 (shown in Figs. 7.13 and 7.15), except we don't need to extract a *Jandata* or a *Plant1*.

Using the Mathcad Add-In in Excel

This method (available in later Mathcad 14 versions) was specially designed for use in Excel; it was also designed for getting data *into* Mathcad *from* Excel and *from* Excel *to* Mathcad. If you want to work primarily in Mathcad, probably you're best option is to use the Excel component, but not a good option if you're working primarily in Excel is to use the Mathcad add-in for Excel.

First, you need to install the add-in.

1. Open Excel with a blank worksheet.

2. Click on *Tools . . . Add-Ins* to get a window like that shown in Fig. 7.23 (yours may be slightly different). This window shows the add-ins that are available on your computer. Most of them are provided as part of Excel: they are Excel's "bells and whistles" that the average Excel user doesn't need so they are not automatically installed when Excel is installed. As a power user, you should definitely check at least *Analysis Toolpak* and *Solver Add-in*! *Mathcad* should be one option available. If it's not, you need to cancel this operation and check the Mathcad Web site for an update. (If you did a Web installation, use the Packaged for the Web program that you downloaded.)

3. Select *Mathcad* and click *OK*.

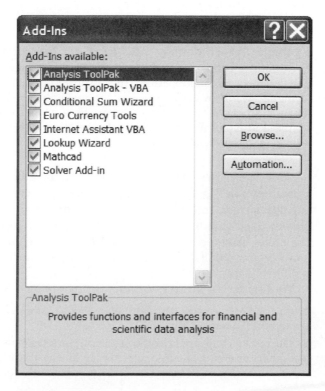

Figure 7.23:
*The Add-Ins window
in Excel*

You should now have a new toolbar in Excel, shown in Fig. 7.24. (It's usually installed on the right border, but you can drag and drop it anywhere you want.) You will also have a new Excel menu item, Mathcad, with the same features as the toolbar. We are most interested in the New Mathcad Object icon, , but the others are very useful too (we'll use some soon).

Figure 7.24:
The Mathcad Toolbar
in Excel

An especially neat icon is the 3D Plot Wizard, ⬛. If you click this icon, Excel will walk you through the simple steps for selecting data on your Excel worksheet, which Mathcad will use to produce a surface plot region on your Excel worksheet! Try it by starting a blank Excel worksheet and generating some data in a set of cells; then click the icon and follow the steps. The big deal about this is that Mathcad's surface plotting is much more sophisticated than Excel's (Try looking for "fog" as a formatting feature in Excel!). (See Appendix: Graphing for a review of Mathcad's graphing bells and whistles.)

With this add-in, we are ready to work with Mathcad within Excel. We demonstrate use of the add-in via a fourth example that is pretty familiar to us by now.

Example 7.4 (Example 7.2, *Kind of Revisited*): Analysis of Production Data An Excel worksheet contains data on the output of a coffee machine manufacturer that has ten production plants. Using a Mathcad add-in in Excel, (a) find the maximum, minimum, mode, and mean of all the plants' production for the year and (b) generate data in Mathcad and plot in Excel a histogram of the production for January for all plants' production.

Once again, to start this example, we need an Excel file similar to that shown in Fig. 7.12. To generate your own version of the file, see the description in the paragraph immediately before Example 7.2. (If you used this file in learning about earlier techniques, such as the Excel Component, you may wish to avoid confusion by creating a new version of Fig. 7.12 (e.g., you can open the file and save under a different name)).

We review the steps necessary for inserting a Mathcad add-in.

1. Within Excel click on the 🟦 icon or click on *Mathcad . . . New . . .* to get the window shown in Fig. 7.25.

2. The options here are fairly obvious: you can either create a new Mathcad worksheet or create one from an existing Mathcad worksheet, and we can insert the add-in into the current, or a new blank, Excel worksheet. In this

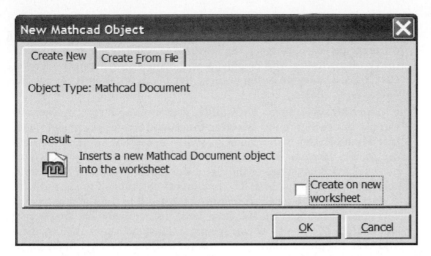

Figure 7.25:
The New Mathcad
Object window in
Excel

example, we will create a new Mathcad worksheet, but the next steps are the same for either choice. After making your choices, click on *OK*.

3. You immediately lose this window and a blank rectangle appears on your Excel worksheet! This can be a bit disconcerting—what do I do now? Actually, the add-in worked: you now have a Mathcad object *embedded* in the Excel worksheet (we stress "embedded" because if you had added in an existing Mathcad worksheet, it would be a *copy*, leaving the original intact). You now need to configure the Mathcad worksheet. Click on the *Set Mathcad Object Properties* icon, 🖻, or use *Mathcad . . . Properties* to get a window similar to that shown in Fig. 7.26. For this window you must select a few things.

    a. *Inputs*: These are the data you want to send *from* Excel *to* the embedded Mathcad region. You must type in the entire range (if it's just one cell, e.g., cell A1, you just type **A1**) of each variable. You can have up to ten of these; Mathcad assigns the variable names $in0, in1, \ldots$.

Figure 7.26:
The Set Mathcad
Object Properties
window in Excel

b. *Outputs*: These are variable values (data) that you wish to send *back to* Excel *from* the embedded Mathcad region. You select only the top-left corner of the Excel range; if the Mathcad variable is a matrix of values, the Excel worksheet will be filled in with values starting at the *Starting Cell*. You can have up to ten of these too; Mathcad assigns the variable names *out0, out1, ...*

c. *Named Ranges*: If you are familiar with naming ranges in Excel, you can use this technique by selecting the named ranges you wish to "map" into Mathcad; they will then be available in drop-down windows in the *Inputs* and *Outputs* frames of Fig. 7.26. To do this step you obviously need to have knowledge of the Excel worksheet layout. In Fig. 7.26 we selected the ranges shown. The two inputs correspond to the entire set of data in the table of the Excel worksheet of Fig. 7.12 and to the first row of data (i.e., just the January data), respectively. The five outputs will contain the results of calculating in Mathcad the max, min, mean, and mode of all the data, the "bins" into which the January data are collected, and the histogram data for January, respectively. This sounds complicated, but all will be made clear! When you are finished with this step, click on *OK*.

4. The rectangular box containing the Mathcad region is still blank; double-click on it (or right-click and choose *Mathcad Object . . . Edit*), and you'll be in it, ready to do some math! (You could also right-click and choose *Mathcad Object . . . Open* to open it in Mathcad itself.) You can now use all the power of Mathcad to perform calculations with inputs *in0, in1, . . .* from Excel, as well as generate new results *out0, out1, . . .* for output to Excel. To perform the calculations shown in Fig. 7.27, let *P* and *Jandata* be the variables representing the entire table and the January data in Fig. 7.12 by setting them equal to the two inputs (this step is not necessary—you could have just worked with variables *in0* and *in1*); then, let the four outputs *out0* through *out3* be the simple statistical computations shown. Define a matrix *out4* and compute a matrix *out5* by using the Mathcad function *hist(out4, Jandata)*, which uses *out4* to define the "bins" into which the values in *Jandata* are counted. The values of *out0* and *out1* will be sent back to Excel. When you are ready to leave Mathcad just click somewhere on the Excel worksheet. You can resize the Mathcad region as you wish to show more or less of the calculations.

$$P := in0 \qquad Jandata := in1$$

**Figure 7.27:
The calculations
in Mathcad for
Example 7.4**

$$out0 := \max(P)$$
$$out1 := \min(P)$$
$$out2 := \text{mean}(P)$$
$$out3 := \text{mode}(P)$$

$$out4 := \begin{pmatrix} 600 \\ 700 \\ 800 \\ 900 \end{pmatrix} \qquad out5 := \text{hist}(out4, Jandata)$$

5. You are not finished yet! To update the Excel worksheet (it is not automatic), click on the *Recalculate Mathcad Objects* icon, ✎, or use menu item *Mathcad . . . Recalculate*. For our example, the final Excel worksheet looks like Fig. 7.28. The results of the Mathcad calculations are outputted to cells E18, E19, E20, and E21, and ranges H19:H22 and I19:I22; we

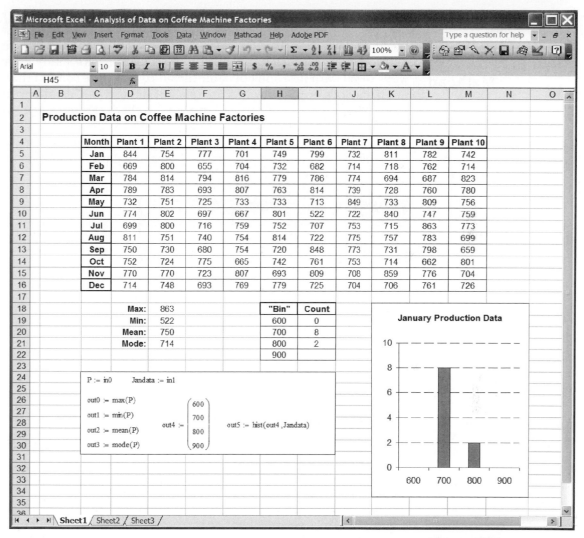

Figure 7.28:
The completed Excel
worksheet with
Mathcad add-in, for
Example 7.4

entered some text and did some formatting of the data in Excel, and then created the bar chart using Excel's *Chart Wizard*.

You can see that using the Mathcad add-in for Excel has some definite potential for organizing and calculating between Excel and Mathcad in a neat, organized fashion. In Example 7.4, we could have actually performed all calculations in Excel (we didn't need Mathcad), but it serves to illustrate the use of the add-in. There *are* calculations that Mathcad can do that either cannot be done with Excel or are awkward to do with it. One obvious candidate is attempting to perform engineering calculations in Excel with data that has a lot of mixed (SI and U.S.) units. Mathcad simplifies this procedure: just output to Mathcad, do the calculations (or simply do the conversions), and input back to Excel!

153

Other Ways to Get Data in and out of Mathcad

We have now discussed in some detail methods for transferring data between Mathcad and Word or Excel. There are many other ways to get data into and out of Mathcad. To illustrate some of these methods, let's suppose we have a small Notepad file, as shown in Fig. 7.29 (make a copy so you can practice the techniques you are about to learn), that we want to do some math with in Mathcad. The data consist of tab-delimited values (meaning that the Tab key, as opposed to, say, commas, was used to separate data in each row).

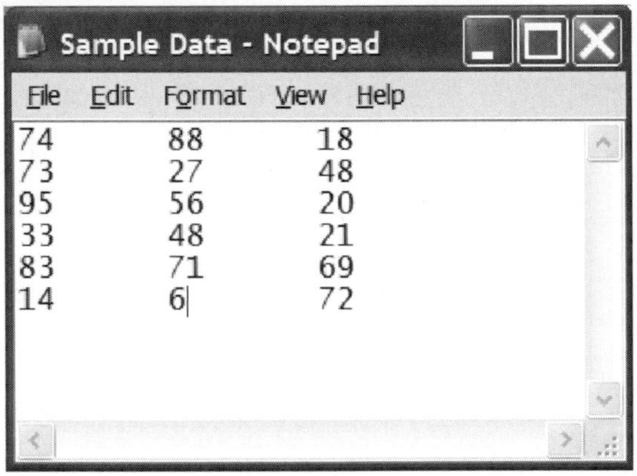

Figure 7.29:
Some sample data

Using the Data Import Wizard

We briefly discussed the Data Import Wizard, shown in Fig. 7.17, earlier. To access this wizard:

1. Click on *Insert . . . Component*, to get the window shown in Fig. 7.16, and choose the *Data Import Wizard* option.

2. Click on *Insert . . . Data . . . Data Import Wizard*, as shown in Fig. 7.30.

This method is a *wizard*, so all you need to do is basically follow the steps; it will walk you through a number of options. The only tricky part is that you need to pay attention to is *File Format*, shown in Fig. 7.17. This refers to how the data are separated in the data file; in Fig. 7.29 both *Fixed Width Text* and *Delimited Text* work to separate the data, but in most data files you would choose *Delimited Text*. After selecting the file and clicking on *Next*, you'll get a window like Fig. 7.31. If everything is working as you'd like, based on the *Preview*, you can either click on *Next* (for more options) or *Finish*; if it's not, you have a choice of trying different delimiters, as shown.

On completion of this wizard you'll have the data displayed in Mathcad in a spreadsheet-type format (or as an icon, depending on your choices with the wizard), with an empty placeholder; in Fig. 7.32 we called this *A*. We're done! Note that this insertion is a *linked* one: if you change the data file, the Mathcad worksheet will

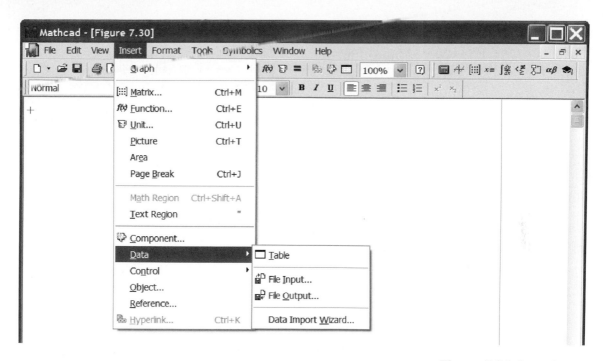

Figure 7.30: Insert Data options

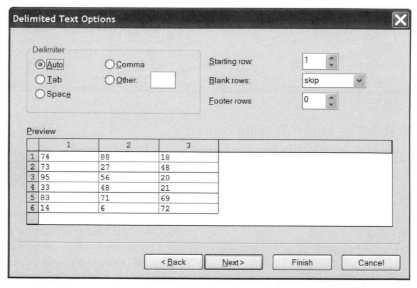

Figure 7.31: Delimit options for data files

be updated the next time you open it or click or double-click on the table. (And if you delete or move the data file you'll end up with an empty table!)

Once you have this data, you can of course analyze the data; you can also right-click on the right side of the equation to change its properties, including changing the data file you are importing.

In Fig. 7.32 we take the data, transpose it, and then save it to a data file using the data transfer technique discussed next.

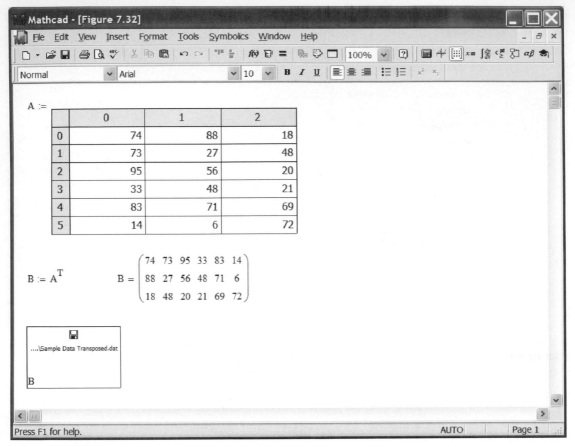

Figure 7.32:
A worksheet
showing receiving
and sending
data

Using File Input and File Output

These methods (one for getting data into Mathcad, the other for data out of Mathcad) are shown as choices in Fig. 7.30. Once again, their use is pretty straightforward. In Fig. 7.32, after computing B, chose *File Output* from Fig. 7.30 and follow the steps to save B's data in a new file (We call it *Sample Data Transposed*, but obviously you can use any name, or even select an existing file—but be careful because it will be overwritten!). The default file extension for output is *.dat; you can tell Windows to open such a file with any of a number of applications, including Notepad and Excel.

In this method, whether you are using File Input or File Output, you'll end up with a floppy disc icon like that in Fig. 7.32, the name of the file, and an empty placeholder ready for you to insert a variable name (we chose to *Show Border* in Fig. 7.32 so you could clearly see the output file). I think it's time Mathcad's graphic was updated here . . . does anyone still use floppies?

Note that clicking *File Output* will update and resave data when you change the Mathcad worksheet, but clicking *File Input* will not—it is a one-time import. If you *do* want to reimport, you can right-click on the File Input region and reselect the data file.

Inserting a Data Table

This method is also shown as a choice in Fig. 7.30, and once again it is pretty straightforward to use. In Fig. 7.33 we show the result of inserting a data table. To complete it you need to type in a variable name at the placeholder, and enter data into the table. You can:

1. Manually type in the data.

2. Copy data from within Mathcad or from another application and select cells to paste to.

3. Right-click on the data table to get access to an *Import* ... option. You can then select a data file to import data from. The data are imported one time only and are no longer connected to the source. You can manually edit the entries as you wish.

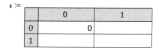

Figure 7.33: A data table

Now that we know how to get data into and out of Mathcad, we can consider some techniques for performing data analysis or statistical analysis. We have already reviewed some basic statistical functions for doing things like finding the max, min, mean, and histogram of data in Examples 7.2 through 7.4.

As you might expect, Mathcad has a plethora of other functions for performing data analysis. We present some of the more commonly needed ones in this section.

Linear Regression

Linear regression is used when you have a set of x, y data (usually with some scatter—perhaps due to experimental or measurement error or uncertainty) that you believe can be represented by a straight line. Figure 7.34 shows an example of this. The data points exhibit some scatter, but we would like to find the *best fit*, or *linear regression*, trend line that best represents the data. The technique we will use is sometimes referred to with those names, but it is also, more formally, known as the *method of least squares*.

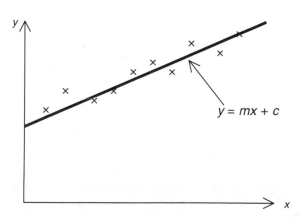

Figure 7.34: Some linear data

We will not discuss the theory behind the method, but we will give a conceptual description. In Fig. 7.34 suppose we stick pins in each of the data points, and hang from each of them loose rubber bands. If we now take a rigid rod and slot it through all these rubber bands, each of them will be in tension—if we pull the rod, say, downwards to the x-axis, all the bands will be trying to pull us back upwards. (In this concept, we assume the pins do not interfere with the rod motion.) If we let the rod go, it will eventually equilibrate (come to rest) at one unique position. This unique position (that shown in Fig. 7.34) is the one position where the total energy stored in the rubber bands is minimized. Because the energy in each rubber band is given by the *square* of its amount of stretch, the unique position is the *least-squared* position!

Obviously, there is some theory involved, which we won't discuss here, in translating this simple concept into useful math. Instead we'll just demonstrate how to use Mathcad to do the work by doing an example.

Example 7.5: Heat Flow An engineer is studying a new insulation material. A cylindrical rod sample of the material (cross-sectional area A = 5 in² and length L = 6 in) is maintained at T_0 = 300 K at one end and is heated at the other to constant temperature T. The heat flow Q in the rod required to maintain the temperature T is measured. The data on heat required for temperature T is shown in the table.

T(K)	350	400	450	500	550	600
Q(W)	2.2	6.8	13	18.5	20	21.5

Find the conductivity k of the material.

The solution to this problem involves the heat conduction equation

$$Q = \frac{kA}{L}(T - T_0)$$

Hence we expect the data for Q vs. T to fall on a straight line

$$Q = \left(\frac{kA}{L}\right)T - \left(\frac{kAT_0}{L}\right)$$

If we can find the slope M (which we use instead of m, to avoid killing our use of "meters" in Mathcad) of the line, we can find k from $k = \frac{ML}{A}$. (In addition, the intercept c should be given by $\frac{kAT_0}{L}$, so we can use this as a check to see if the data are good!)

Mathcad has a number of built-in functions that automate linear regression. Some of the more useful functions (there are others) are:

1. *slope*(*x*, *y*): This function computes the slope of the line for the least-squares fit between (vector) data *x* and (vector) data *y*. Obviously, *x* and *y* must be of the same size.

2. *intercept*(*x*, *y*): This function computes the intercept of the line for the least-squares fit between *x* and *y*.

3. *corr*(*x*, *y*): This function computes the correlation coefficient of the least-squares fit (a measure of how well the data fit the line).

We can immediately solve the problem. Figure 7.35 shows the complete solution. We can see that the data are pretty good (the correlation coefficient is close to unity, which is good, and also $c \approx \dfrac{kAT_0}{L}$). We need to be careful in this worksheet by:

1. Typing the data for *T* and *Q* into two matrices, but any of the data import methods we discussed in this chapter could have been used.
2. Using a literal subscript for T_0.
3. Defining a new function of *T*, $Q_{LS}(T)$, as shown.
4. Preparing a nicely formatted graph. (See Appendix: Graphing for help.)

Specialized (Nonlinear) Regression

Not all experimental data are linear. Some common nonlinear trends Mathcad can handle are (assuming *x* and *y* are arrays of data):

1. *Exponential*: For example, the temperature of a cooling body is exponential with time. For this we can use the built-in function

expfit(*x*, *y*, [*k*])

This computes a 3×1 matrix containing the best values of *a*, *b*, and *c* for equation

$$y = a \cdot e^{bx} + c$$

In this function, *k* is an optional matrix containing guess values for *a*, *b*, and *c*.

2. *Logarithmic*: For this we can use *either* the built-in function

lnfit(*x*, *y*)

This computes a 2×1 matrix containing the best values of *a* and *b* for equation

$$y = a \cdot \ln(x) + b$$

or the built-in function

logfit(*x*, *y*, *k*)

This computes a 3×1 matrix containing the best values of *a*, *b*, and *c* for equation

$$y = a \cdot \ln(x + b) + c$$

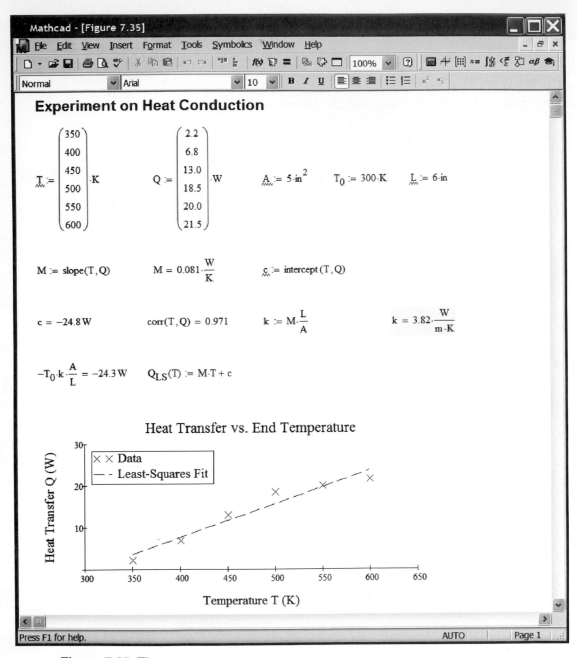

Figure 7.35: The worksheet for Example 7.5

In this function, k is a matrix containing guess values for a, b, and c.

3. *Power*: For this we can use the built-in function

$$pwrfit(x, y, k)$$

This computes a 3×1 matrix containing the best values of a, $b,$ and c for equation

$$y = a \cdot x^b + c$$

In this function, k is a matrix containing guess values for a, b, and c.

4. *Sine*: For this we can use the built-in function

$sinfit(x, y, k)$

This computes a 3×1 matrix containing the best values of a, b, and c for equation

$y = a \cdot \sin(x + b) + c$

In this function, k is a matrix containing guess values for a, b, and c.

5. *Logistic*: For this we can use the built-in function

$lgsfit(x, y, k)$

This computes a 3×1 matrix containing the best values of a, b, and c for equation

$$y = \frac{a}{1 + be^{-cx}}$$

In this function, k is a matrix containing guess values for a, b, and c.

Since each of these works in the same way, we illustrate using only one of them.

Example 7.6: Oil Viscosity An engineer is testing new motor oil for its viscosity. She needs to know whether it will be viscous enough when the engine temperature exceeds 280°F and has the following data available (notice the mixed units):

T (°F)	100	120	150	200	220	240
μ(N·s/m²)	1.6	1.3	1.0	0.56	0.50	0.37

The oil must have a viscosity of at least 0.20 N·s/m² at 280°F. Will the oil be okay?

It turns out that the viscosity of oil has an exponential dependence on the temperature

$\mu(T) = ae^{bT} + c$

We need to use the data to find the best values of a, b, and c, and then we can *extrapolate* using the equation to find an estimate for μ at $T = 280$°F.

The solution is shown in Fig. 7.36. We used the function expfit to find the best values of a, b, and c, and then defined a new function $\mu_{\text{Fit}}(T)$ with which we estimated μ at $T = 280$°F. It looks like the oil is okay (although we should always be cautious when extrapolating, as opposed to interpolating, in which we would estimate new values *within* the given temperature range). Note that we:

1. Typed the data for T and μ into two matrices, but we could use any of the data import methods discussed in this chapter.

2. Did not use units. We *cannot* use units with these methods.

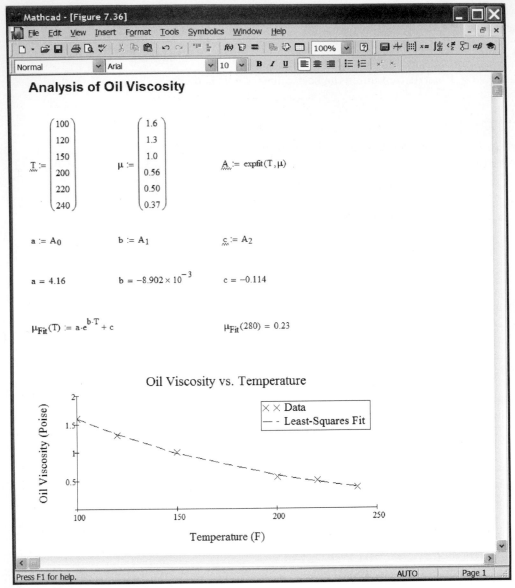

Analysis of Oil Viscosity

$$\underset{\sim}{T} := \begin{pmatrix} 100 \\ 120 \\ 150 \\ 200 \\ 220 \\ 240 \end{pmatrix} \qquad \mu := \begin{pmatrix} 1.6 \\ 1.3 \\ 1.0 \\ 0.56 \\ 0.50 \\ 0.37 \end{pmatrix} \qquad \underset{\sim}{A} := \text{expfit}(T,\mu)$$

$$a := A_0 \qquad\qquad b := A_1 \qquad\qquad \underset{\sim}{c} := A_2$$

$$a = 4.16 \qquad\qquad b = -8.902 \times 10^{-3} \qquad c = -0.114$$

$$\mu_{\text{Fit}}(T) := a \cdot e^{b \cdot T} + c \qquad\qquad \mu_{\text{Fit}}(280) = 0.23$$

Oil Viscosity vs. Temperature

Figure 7.36: *The worksheet for Example 7.6*

3. Use array subscripts to extract a, b, and c from A and a literal subscript for μ_{Fit}.

4. Defined a new function of T, $\mu_{\text{Fit}}(T)$, as shown.

5. Prepared a nicely formatted graph. (See Appendix: Graphing for help.)

Polynomial Fit

This method can be used to fit data to a polynomial of the form

$$Y = a_0 + a_1X + a_2X^2 + a_3X^3 + \ldots + a_nX^n$$

and involves two built-in functions:

1. *regress*(X, Y, n): This function produces a vector of values, which *interp* (below) uses to find the nth order polynomial that best fits the X and Y data values.

2. *interp*(V, X, Y, x): This function computes an interpolated y value corresponding to x, using the output vector V from the *regress* function.

We use capital letters here (but you can use your own notation) to distinguish data X and Y from computed estimates y at points x.

Figure 7.37 on page 164 shows an example of this method; you should try to reproduce it. We can make some comments:

1. We fitted a third-order polynomial to the X, Y data, so n = 3.

2. We evaluated the special V vector, even though it's not very useful: only the last n + 1 (or, in this example, four) terms have much meaning for us; they are the n + 1 a coefficients in the polynomial equation.

$$X := \begin{pmatrix} 1 \\ 2 \\ 3.5 \\ 4.1 \\ 5.2 \\ 5.7 \end{pmatrix} \qquad Y := \begin{pmatrix} 0.83 \\ 1.4 \\ 2.2 \\ 2.7 \\ 3.7 \\ 4.4 \end{pmatrix}$$

$$V := regress\,(X,Y,3) \qquad y(x) := interp\,(V,X,Y,x) \qquad V = \begin{pmatrix} 3 \\ 3 \\ 3 \\ 0.08 \\ 0.894 \\ -0.168 \\ 0.025 \end{pmatrix} \qquad \overrightarrow{y(X)} = \begin{pmatrix} 0.83 \\ 1.4 \\ 2.23 \\ 2.66 \\ 3.73 \\ 4.39 \end{pmatrix}$$

$$x := 0, 0.01 .. 6$$

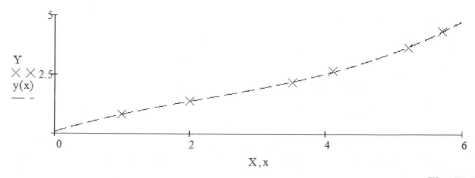

3. We used interp to define $y(x)$, a function of x.

4. We needed to vectorize (by using the ⟨⟨⟩⟩ icon on the Vector and Matrix Toolbar or by typing **Ctrl + -**) the evaluation of y at the original X data points.

5. Plotting a graph that includes a *vector of values* and a *function* can be tricky. (See Appendix: Graphing for help.)

Other Regressions, Splines, and More

The methods described above should suffice for most purposes, but if you still haven't had enough, consider the following:

1. *genfit*: This function is a regression for fitting data to an *arbitrary* function. For example, it can be used to find a and b in $y = a \cdot x \cdot e^{b \cdot x}$.

2. *linfit*: This function is a regression for fitting a *linear* combination of *arbitrary* functions to data. For example, it can be used to find a, b, and c in $y = a \cdot x + b \cdot e^{2x} + c \cdot \sin(x^2)$.

3. Mathcad also has built-in capabilities for obtaining *splines* of various kinds.

For help on these and lots of other numerical analysis features search Mathcad's Help for Data Analysis to get the window shown in Fig. 7.38. (Also see Exercises 7.22 and 7.23.)

Exercises

Note: Be aware that you may encounter problems if you solve a problem that involves a variable (say, x), and then use the same variable in a later problem in the same worksheet. If you get some strange results, this may be the reason; if so, make sure you reset the variable to a null value (e.g., set $x = 0$), or simply solve those problems on separate worksheets.

7.1 A breakfast cereal manufacturer must maintain a certain minimum weight of each box. He has a sample of data for the box weights W (kg) from the production line as follows:

W (kg)				
1.53	1.46	1.55	1.48	1.50
1.52	1.55	1.46	1.5	1.57
1.55	1.63	1.48	1.44	1.53
1.52	1.45	1.52	1.58	1.51

Find the max, min, and mean of this data. Then plot a histogram and determine how many boxes are (a) below 1.4 kg and (b) below 1.45 kg.

7.2 A farmer wishes to estimate the condition of her herd of cattle based on data on cattle weight. The cattle weight data are as follows:

W (kg)					
1866	1934	1898	2206	1880	1831
1677	1975	2252	2018	1927	1999
1843	1599	2243	2217	1454	2177

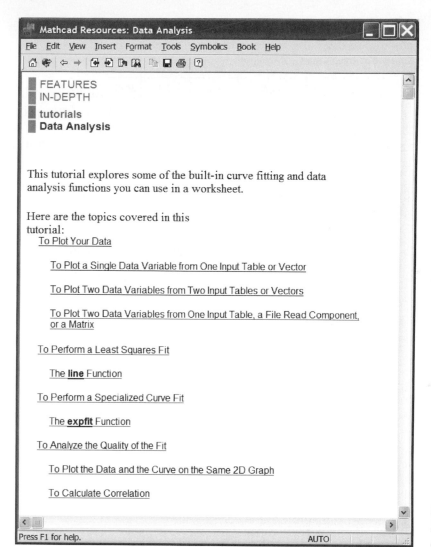

Find the max, min, and mean of this data. Then plot a histogram and determine how many cattle are (a) below 1600 kg, (b) below 1800 kg, and (c) at least 2000 kg.

7.3 The grades for a course Introduction to Mathcad are as follows:

Grades (%)	73	84	77	81	98	90
	79	92	85	96	77	94
	73	98	80	100	76	69
	73	85	100	81	89	93

Find the max, min, and average grade. Then plot a histogram and determine how many students failed the course (to pass you need at least 60%).

7.4 Find the slope and intercept of the best straight-line fit to the following data:

x	0	1	2	3	4	5	6	7	8	9	10
y	−1.91	−0.46	0.89	2.03	3.65	5.45	5.61	7.19	9.40	9.77	11.8

Estimate the values of (a) $y(2.5)$, (b) $y(5.5)$, (c) $y(11)$, (d) $x(y = 5)$, (e) $x(y = 13)$, and indicate the goodness of fit of the data to the line.

7.5 Find the slope and intercept of the best straight-line fit to the following data:

x	0	1	2	3	4	5	6	7	8	9	10
y	3.30	3.11	2.10	1.66	0.96	0.24	−0.48	−1.24	−1.77	−2.39	−3.32

Estimate the values of (a) $y(3.5)$, (b) $y(6.5)$, (c) $y(-1)$, (d) $x(y = 4)$, (e) $x(y = -1)$, and indicate the goodness of fit of the data to the line.

7.6 Find the slope and intercept of the best straight-line fit to the following data:

x	0	1	2	3	4	5	6	7	8	9	10
y	2.74	4.95	6.22	8.37	10.5	13.6	13.6	15.6	19.1	20.5	21.5

Estimate the values of (a) $y(1.5)$, (b) $y(8.5)$, (c) $y(-1)$, (d) $x(y = 10)$, (e) $x(y = 24)$, and indicate the goodness of fit of the data to the line.

7.7 An engineer is testing a new material to find its modulus of rigidity E. The change in length δ of a rod of the material (length $L = 4$ m, diameter $d = 1$ cm) due to applied force F is measured for a range of forces, as shown in the table. Estimate the best value of E from the data.

F (kN)	2	4	6	8	10	12	14	16	18	20
δ(mm)	0.82	1.47	2.05	3.37	3.75	4.17	5.25	5.44	6.62	7.97

Figure E7.7

Note that $E = \dfrac{\sigma}{\epsilon} = \dfrac{\left(\dfrac{W}{\pi d^2 / 4}\right)}{\left(\dfrac{\delta}{L}\right)} = \left(\dfrac{4L}{\pi d^2}\right)\left(\dfrac{W}{\delta}\right)$, where σ and ϵ are the stress and strain, respectively. Instead of using this equation to get several values of E from *individual* pairs of W and δ, find the best estimate of $\left(\dfrac{W}{\delta}\right)$ from the slope of the W vs. δ line!

7.8 A bicycle manufacturer wants to determine how successful its advertising budget is in increasing sales. Based on the data below, do the sales increase as the advertising budget does—is there a strong correlation between the two? If so, use the data to predict the sales expected if the advertising budget is increased to $35,000 for the eighth quarter. Also find the anticipated sales if there was *no* advertising.

Quarter	1st	2nd	3rd	4th	5th	6th	7th
Advertising Budget (× $1000)	20	15	22	31	18	15	27
Bicycle Sales (× 1000)	5.5	4.5	5.5	7	5	5	6.5

7.9 Consider the following data:

x	0	1	2	3	4	5	6	7	8	9	10
y	8.11	4.98	4.14	3.22	2.89	2.39	2.24	2.16	2.09	2.08	2.05

Find the best values of a, b, and c in $y = a \cdot e^{bx} + c$ that fit the data. Plot the data and equation on a suitable graph (e.g., log-normal, log-log). Estimate $y(12)$ and $x(y = 5)$.

7.10 Consider the following data:

x	−5	−4	−3	−2	−1	0	1	2	3	4	5
y	1. 68	1.64	1.48	1.25	0.73	0.41	−0.12	−0.87	−3.33	−5.21	−6.05

Find the best values of a, b, and c in $y = a \cdot e^{bx} + c$ that fit the data. Plot the data and equation on a suitable graph (e.g., log-normal, log-log). Estimate $y(6)$ and $x(y = 1.5)$.

7.11 In an experiment the cooling of a body in a quenching fluid is to be measured. Theory says that the temperature T should behave like $T = T_0 e^{-\frac{t}{\tau}}$, where τ is the time constant for the process. The value of τ provides information about the convection and conduction taking place. The following data were measured:

t (s)	0	5	10	15	20	25	30
T (°C)	235	215	127	96	70	40	31

Estimate a value for τ. Plot the data and equation on a suitable graph (e.g., log-normal, log-log).

7.12 Consider the following data:

x	1	2	3	4	5	6	7	8	9	10
y	−0.08	1.05	1.69	2.32	2.79	2.80	3.21	3.99	4.16	4.28

Find the best values of a and b in $y = a \cdot \ln(x) + b$ that fit the data. Plot the data and equation on a suitable graph (e.g., log-normal, log-log). Estimate $y(12)$ and $x(y = 4.5)$.

7.13 Consider the following data:

x	1	2	3	4	5	6	7	8	9	10
y	5.00	3.82	2.90	2.36	2.17	1.45	1.89	1.13	1.01	0.65

Find the best values of a and b in $y = a \cdot \ln(x) + b$ that fit the data. Plot the data and equation on a suitable graph (e.g., log-normal, log-log). Estimate $y(12)$ and $x(y = 2.5)$.

7.14 Consider the following data:

x	1	1.5	2	2.5	3	3.5	4	4.5	5
y	1.36	1.78	2.38	3.27	4.12	4.88	6.01	6.54	8.02

Find the best values of a, b, and c in $y = a \cdot x^b + c$ that fit the data. Plot the data and equation on a suitable graph (e.g., log-normal, log-log). Estimate $y(5.5)$ and $x(y = 5)$.

7.15 Consider the following data:

x	1	1.5	2	2.5	3	3.5	4	4.5	5
y	3.39	1.86	1.27	1.06	1.18	1.06	0.93	1.03	0.87

Find the best values of a, b, and c in $y = a \cdot x^b + c$ that fit the data. Plot the data and equation on a suitable graph (e.g., log-normal, log-log). Estimate $y(5.5)$ and $x(y = 0.5)$.

7.16 In tests on the horsepower P required to cruise at speed V (mph) of a hybrid vehicle, the following data were obtained:

V (mph)	10	20	30	40	50	60
P (hp)	8.60	11.8	19.6	37.6	68.9	102

Theory indicates that the relation between P and V should be $P = a \cdot V^b + c$. Find the best values of a, b, and c that fit the data. Plot the data and equation on a suitable graph (e.g., log-normal, log-log). Estimate the horsepower required to go at 75 mph, and the speed attained when horsepower is 55 hp.

7.17 Consider the following data:

x	0	$\pi/4$	$\pi/2$	$3\pi/4$	π	$5\pi/4$	$3\pi/2$	$7\pi/4$	2π
y	5.68	4.88	2.82	0.71	0.00	0.87	2.51	4.83	5.35

Find the best values of a, b, and c in $y = a \cdot \sin(x + b) + c$ that fit the data. Plot the data and equation.

7.18 Consider the following data:

x	0	1	2	3	4	5	6	7	8
y	0.99	1.53	1.88	1.99	2.62	2.57	2.44	2.51	2.76

Find the best values of a, b, and c in $y = \dfrac{a}{1 + be^{-cx}}$ that fit the data. Plot the data and equation. Estimate $y(10)$ and $x(y = 2.5)$.

7.19 Consider the following data:

x	0	1	2	3	4	5	6	7	8
y	1.37	2.39	3.27	4.50	4.37	4.57	4.55	4.72	4.92

Find the best values of a, b, and c in $y = \dfrac{a}{1 + be^{-cx}}$ that fit the data. Plot the data and equation. Estimate $y(10)$ and $x(y = 4)$.

7.20 Consider the following data:

x	0	1	2	3	4	5	6	7	8
y	13	14	15	11	6	0	−8	−19	−27

Find the best values of a, b, and c in $y = a + bx + cx^2$ that fit the data. Plot the data and equation. Estimate $y(-1)$ and $x(y = 4.5)$.

7.21 Consider the following data:

x	−5	−4	−3	−2	−1	0	1	2	3	4	5
y	−16	−9	−4	−1	0	0	1	1	2	5	8

Find the best values of a, b, and c in $y = a + bx + cx^2 + dx^3$ that fit the data. Plot the data and equation. Estimate $y(4.5)$ and $x(y = 4.5)$.

7.22 Consider the following data for the concentration C of a medication in the bloodstream applied to the surface of a patient's skin:

t (hr)	0.2	0.4	0.6	0.8	1
C (μm)	0.9	1.3	1.5	1.7	1.8

It is believed that the following model can be applied: $C(t) = a_0 te^{-3t} + a_1\sqrt{t}$. Find the best values of a_0, and a_1 that fit the data. Plot the data and equation. Estimate the concentration after $1\frac{1}{2}$ hr, and the time at which the concentration reaches 2 μm.

Notes: You need to use *linfit* here. The steps are:

a. Define two column vectors t and C containing the data on t and C.

b. Define the two functions that you need to assemble (see Fig. E7.22a).

c. Define a column vector $F(t)$ containing all the functions used (see Fig. E7.22b). Note that you can combine steps 1 and 2, and define $F(t)$ as in Fig. E7.22c.

d. Generate a_0 and a_1 by using the built-in function linfit (see Fig. E7.22d), and create a new function $C_{Fit}(t)$.

Figure E7.22a $f(t) := t \cdot e^{-3 \cdot t}$ $\qquad$ $g(t) := \sqrt{t}$

Figure E7.22b $F(t) := \begin{pmatrix} f(t) \\ g(t) \end{pmatrix}$

Figure E7.22c $F(t) := \begin{pmatrix} t \cdot e^{-3 \cdot t} \\ \sqrt{t} \end{pmatrix}$

$a := linfit(t, C, F)$

Figure E7.22d $C_{Fit}(t) := a_0 \cdot f(t) + a_1 \cdot g(t)$

Note: The function linfit must be used as illustrated: *don't* type in $F(t)$, for example! This creates a matrix of coefficients (in this case a). Assuming *ORIGIN* is set to the default value of zero, the elements of a will be, in this case a_0, and a_1, with *array subscripts*!

Problem solved! To find the time at which the concentration reaches 2 μm, one option is to use Given . . . Find.

7.23 Consider the following data:

x	0	1	2	3	4	5	6	7	8	9	10
y	6	7	5	2	-2	-3	-1	2	3	1	-2

Find the best values of a_0, a_1, and a_2 in $y(x) = a_0 \sin(x) + \dfrac{a_1}{1 + x} + a_2 e^{-\frac{x}{2}}$

that fit the data. Plot the data and the equation. Estimate $y(12)$, and $x(y = 0)$.

Note: Use the hints provided in Exercise 7.22 as a guide.

Appendix: Graphing

In this appendix we review most of the ways to graph functions and data in Mathcad (and ways that graphing gets messed up). Each section is basically self-contained. You may go directly to a section to find out how to accomplish specific graphing tasks.

A.1

Accessing Mathcad's Graphing Features

To gain access to Mathcad's graphing features, you can:

1. Select menu item *Insert . . . Graph . . .* to get options shown in Fig. A.1.

2. Click on the Graph Toolbar icon, 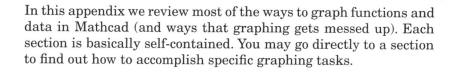, on the Math Toolbar shown in Fig. A.2, to get the Graph Toolbar shown in Fig. A.3.

3. Type an accelerator key for quick access to a feature (e.g., type @ in an empty region to get a blank X-Y Plot). Several accelerator keys are shown in Fig. A1.

A.2

X-Y Plots

These two-dimensional graphs are probably the most common graphs you'll need to create. You might want to plot one of the following:

1. One or more functions—for example, $f(x) = x \sin(x)$ and/or $g(x) = e^{-\frac{x}{5}} \cos(2x)$ vs. x.

2. One or more sets of data—for example, a set of y values and/or a set of z values against a set of x values.

3. A combination of plots 1 and 2—for example, a function $f(x)$ and a set of y values against a set of x values.

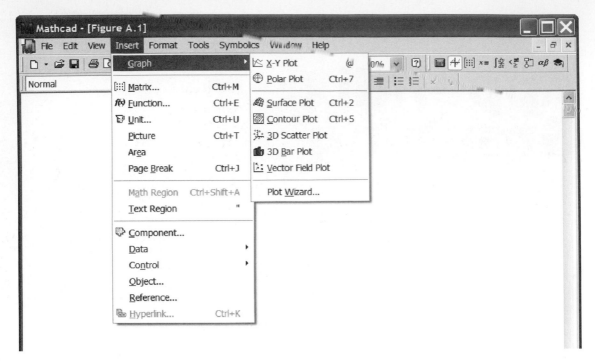

Figure A.1
The Insert Graph
Menu

Figure A.2
The Math Toolbar

Figure A.3 The
Graph Toolbar

We will demonstrate these plots using functions such as

$$f(x) = x \sin(x)$$

$$g(x) = e^{-\frac{x}{5}} \cos(2x) \cdot \Phi(x - 1)$$

$$h(t) = t - \frac{t^2}{10} + \frac{1}{t - 5}$$

$$w(x, y) = \sin(y)\tan(x)$$

These were chosen because they have some interesting features: $g(x)$ has a Heaviside step function (to get it type **F** and then immediately **Ctrl + G**, or use the Greek Symbol Toolbar); $h(t)$ is a function of t, not x (so we can see what happens when we plot, say, g_2 and h_3 on the same graph), and it blows up at $t = 5$; and $w(x,y)$ is a function of two variables (and it also blows up when $x = x = \frac{\pi}{2}, \frac{3\pi}{2}, \frac{5\pi}{2}, \ldots$).

Plotting One or More Functions Using QuickPlot

In this method Mathcad automatically sets up the graph so that the plot goes from –10 to 10. You simply create an X-Y Plot by doing any of the following (in order of convenience):

172

1. Type @.
2. Click the X-Y Plot icon, 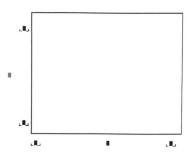 , on the Graph Toolbar.
3. Use *Insert . . . Graph . . . X-Y Plot*.

You will obtain a blank graph, as shown in Fig.A.4, with six placeholders. For now we're only interested in the two *required* ones (▪), and not the *optional* ones (▪.). In the required placeholder on the vertical axis, we can type one of the functions listed above; in the required placeholder on the horizontal axis, we must type the independent variable.

Figure A.4 The blank
X-Y Plot

Figure A.5 show the results of defining the functions and using QuickPlot. We can make a few comments here:

1. We plotted functions, but you could just as well type in the expression you wish to plot. For example, instead of $f(x)$, we could plot $x \sin(x)$.

2. The function $f(x)$ plotted just fine.

3. The function $g(x)$ also plotted just fine—note that the Heaviside step function did its job!

4. Although we defined $h(t)$ as a function of t, we can *evaluate* and *plot h* with any argument we choose; we can evaluate, say, $h(5)$, or, as in this case, plot $h(x)$ vs. x.

5. The function $w(x, y)$ did *not* plot. When we ask QuickPlot to graph $w(x, y)$, it automatically evaluates $w(x, y)$ for a range of x values, but it needs to be told the value of y! To remedy this error, we need to specify y *before* the graph is plotted and computed. (See Fig. A.6 for an example.)

6. These graphs need formatting. We leave the details of this to a later section, but if you want to play, just double-click on one of the graphs!

7. Even without formatting, we can tweak the appearance of the graphs. We can now click on each graph, and then click on the optional placeholders ( ), and type in desired ranges for the *dependent* and *independent* ranges. For example, we did the following to get the graphs shown in Fig. A.7 :

   a. Double-clicked on each graph to get the Formatting Currently Selected X-Y Plot window and changed the axes style to *crossed*.

   b. For the $f(x)$ graph, we typed in **0** and **3*p Ctrl + G** (to get 3π; alternatively, to get π you can use the Greek Symbol Toolbar) in the two placeholders on the horizontal axis. Note that you don't need to delete the values Mathcad placed in these locations—you can just type in the new values.

c. For the $g(x)$ graph, we typed in **0** and **5*p Ctrl + G** in the two placeholders on the horizontal axis.

d. For the $h(x)$ graph, we typed in **−5** and **20** in the two placeholders on the horizontal axis, and **−5** and **5** in the two placeholders on the vertical axis. We typed in values on the vertical axis to not have large values "swamp" the vertical scale; *this is a very useful technique when you have a function that has a region where the function value becomes very large.*

e. For the $w(x, y)$ graph, we typed in **−p Ctrl + G** and **p Ctrl + G** in the two placeholders on the horizontal axis, and **−1** and **1** in the two placeholders on the vertical axis (as well as setting y to 3).

$$f(x) := x \cdot \sin(x) \qquad g(x) := e^{-\frac{x}{5}} \cdot \cos(2 \cdot x) \cdot \Phi(x-1) \qquad h(t) := t - \frac{t^2}{10} + \frac{1}{t-5} \qquad w(x,y) := \sin(y) \cdot \tan(x)$$

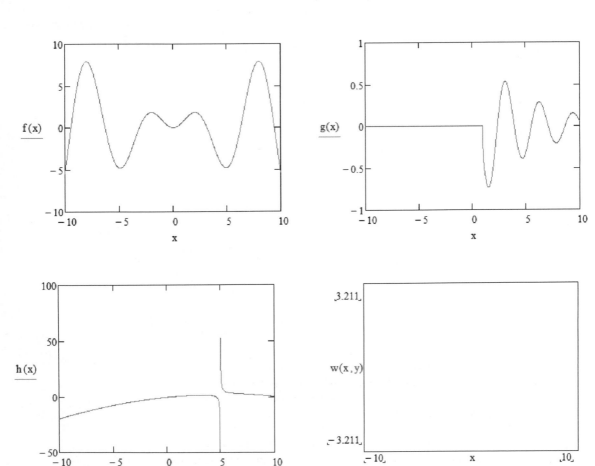

Figure A.5 Some Quick Plot examples

$y := 3$

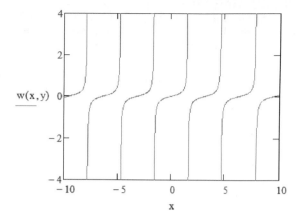

Figure A.6
The function w(x, y)
for y = 3

You can see that QuickPlot is quite simple to use. Probably the most common error in using it is using an independent variable (in our case, x) that has previously been defined somewhere else on the worksheet. If you defined the independent variable with the specific intent of defining a graphing range (see the next section), then there's not usually a problem. On the other hand, look what kind of QuickPlot we get for $f(x)$ and $g(x)$ if we define $x = 2$ (Fig. A.8), or $x = 2, 3, 4$ (Fig. A.9—to get the x range type **x:2;4**). Not exactly what we had in mind!

To plot several functions on the same graph, we simply type each function in the placeholder on the vertical axis *separated by a comma*. In Fig. A.10 we created a QuickPlot of $f(x)$ and $g(x)$. It seems that $f(x)$ gets "chopped" at large values—and it pretty much swamps $g(x)$. To modify the display, you can type values into placeholders as described in item 7 above. Note that Mathcad assigns default trace types; to modify these see the later section on formatting.

Some points to bear in mind when plotting more than one function using QuickPlot are:

1. You can plot up to 16 functions on one graph.

2. QuickPlot is designed to provide a quick plot of one or more functions of one variable; that is, the *same* one independent variable (in all the graphs so far, this was x). If you wish to plot several functions that have differing independent variables (e.g., suppose you want to plot $f(x)$ vs. x and $h(t)$ vs. t for different ranges of x and t) on the same graph, see the next section.

3. You can plot two functions *parametrically* (i.e., you can plot one against the other and not against the independent variable). Figure A.11 shows a classic example of this, a Lissajous figure generated by the two functions shown.

$$f(x) := x \cdot \sin(x) \qquad g(x) := e^{-\frac{x}{5}} \cdot \cos(2 \cdot x) \cdot \Phi(x-1) \qquad h(t) := t - \frac{t^2}{10} + \frac{1}{t-5} \qquad w(x,y) := \sin(y) \cdot \tan(x)$$

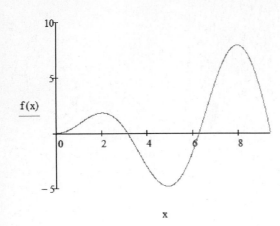

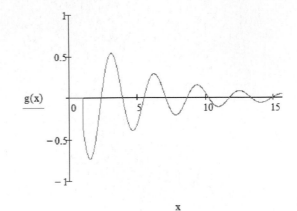

$$y := 3$$

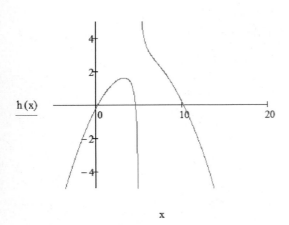

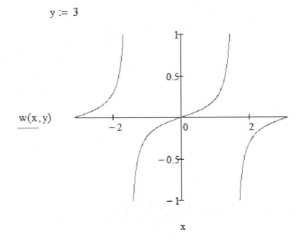

Figure A.7
The QuickPlots after
"tweaking"

There is one other feature we need to look at—using *units* in QuickPlots.

Figure A.12 shows an example of using units with a QuickPlot graph. We defined an initial upward velocity $v_0 = 100$ ft/s of a ball (use a literal subscript; i.e., type **v.0(t)**), and defined the equation for the height of the ball y as a function of time t (note that acceleration of gravity g is built in so it doesn't need to be defined, and that we ignore aerodynamic drag). The maximum height y_{max} of the ball is also defined and evaluated (and in the evaluation of y_{max} we clicked on the units placeholder and typed **ft**; we also formatted the answer to zero decimal places).

$$f(x) := x \cdot \sin(x) \qquad g(x) := e^{-\frac{x}{5}} \cdot \cos(2 \cdot x) \cdot \Phi(x-1) \qquad h(t) := t - \frac{t^2}{10} + \frac{1}{t-5} \qquad w(x,y) := \sin(y) \cdot \tan(x)$$

$$x := 2$$

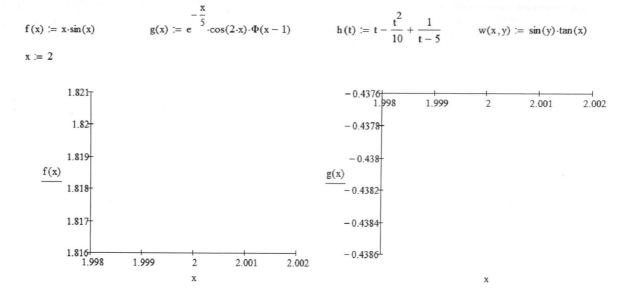

Figure A.8
Plots with x = 2

$$f(x) := x \cdot \sin(x) \qquad g(x) := e^{-\frac{x}{5}} \cdot \cos(2 \cdot x) \cdot \Phi(x-1) \qquad h(t) := t - \frac{t^2}{10} + \frac{1}{t-5} \qquad w(x,y) := \sin(y) \cdot \tan(x)$$

$$x := 2 .. 4$$

Figure A.9
Plots with x = 2, 3, 4

The QuickPlot graph is not exactly what we want: it plots for negative as well as positive times! We can fix this by typing **0** and, say, **5** in the two optional placeholders on the horizontal axis and get the graph shown in Fig. A.13. Now we have a new problem: it has numbers on the vertical axis that don't correspond with the maximum value of 155 ft!

**Appendix:
Graphing**

$$f(x) := x \cdot \sin(x) \qquad\qquad g(x) := e^{-\frac{x}{5}} \cdot \cos(2 \cdot x) \cdot \Phi(x - 1)$$

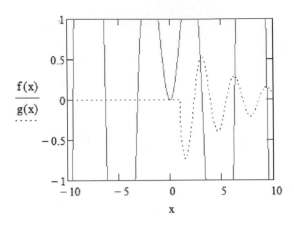

**Figure A.10
A QuickPlot of two
functions**

$$x(t) := \sin(2 \cdot t) \qquad y(t) := \cos(5 \cdot t)$$

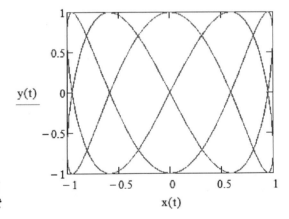

**Figure A.11
A parametric plot**

This example illustrates an important point that leads to some confusion when using units in a graph: *Mathcad uses scales on the horizontal and vertical axes that are based on the default units for the worksheet.* The default units in Mathcad are SI, so the vertical scale is in meters. If we were plotting, say, power, the numbering would correspond to watts. You can change the default units to, for example, U.S. units, by clicking *Tools ... Worksheet Options ... Unit System* to get the window shown in Fig. A.14. Hence, to make a graph appear with numbering that corresponds to your calculation units it is often a matter of choosing the right unit system for your worksheet. In fact, there are two ways to make your graphs with units to display as you'd like them:

$$v_0 := 100 \cdot \frac{ft}{s} \qquad y(t) := v_0 \cdot t - \frac{1}{2} \cdot g \cdot t^2 \qquad y_{max} := \frac{v_0^2}{2 \cdot g} \qquad y_{max} = 155 ft$$

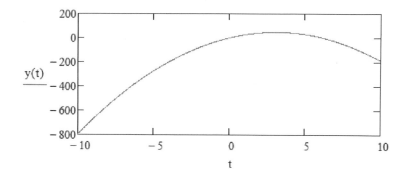

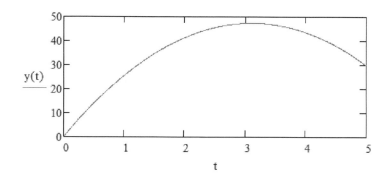

**Figure A.12
A QuickPlot with units**

**Figure A.13
The QuickPlot after
"tweaking"**

1. Choose the appropriate unit system using the window shown in Fig. A.14.

2. Instead of plotting the variables, plot each one divided by the units in which you would like the graph numbering to appear. This divides the variable by those units, leaving a dimensionless plot.

The second approach is illustrated in Fig. A.15. We divided by ft; the horizontal axis was left alone because the default unit *is* seconds (if we wanted a scale of hours, we would have to use t/hr on the horizontal axis). Everything looks consistent now! The formatting leaves something to be desired; we discuss graph formatting in a later section.

Plotting One or More Functions by Defining a Range or Ranges

In this method, before inserting a graph, you *define a range of values for the independent variable*. For example, to define a range x varying from a to b in Mathcad, you can:

*Figure A.14
Selecting default
units*

$$v_0 := 100 \cdot \frac{ft}{s} \qquad y(t) := v_0 \cdot t - \frac{1}{2} \cdot g \cdot t^2 \qquad y_{max} := \frac{v_0^{\,2}}{2 \cdot g} \qquad y_{max} = 155\,ft$$

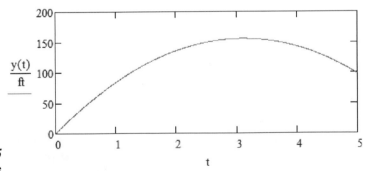

*Figure A.15
The plot with units
fixed*

1. Type **x:a;b** where a and b are the beginning and end of the range (they must be real, but can be integers, decimals, or fractions, or previously defined variables). Mathcad replaces the semicolon (;) with two dots (..); *you cannot type the two dots!* (Instead of typing **;** you can get them by using the ▣ icon on the Vector and Matrix Toolbar). Regardless of the values of a and b, Mathcad fills in the range between the two *using a step size of unity* (e.g., for an increasing range, $a, a + 1, a + 2$, until we reach b); the last value will be less than or equal to b, if a and b are decimals.

2. Type **x:a,b;c** where *a* is the beginning value, *b* is the second value in the series, and *c* is the end of the range (with the same constraints as in 1 above). Mathcad again replaces the semicolon (;) with two dots (..). *The most common error in using this approach is to assume that the middle value, b, represents the step size rather than the second value!* Note that if you wish to define a range with units, you *must* use this three-number format, and you obviously must use consistent dimensions with the three values.

Figure A.16 shows an example of using a range. It's quite ugly but serves to illustrate probably the most common mistake in using this graphing method—using method 1 above for defining the range. With that method, the step size is unity (even if you are using units, e.g., the step size would be 1 m, or 1 s). It's almost always better to use the second method; compare Figs. A.16 and A.17! The only situation in which you'd prefer method 1 for defining your graphing range is when you're plotting a bar graph (see the later section on formatting).

Unlike the QuickPlot method, the horizontal scale in this method is defined by the independent variable range you defined; if you wish, you can still set upper and lower limits on the horizontal axis. With this one exception, most of the discussion in the previous section applies. You can review that section for information on:

1. Scaling the vertical axis. Note that unlike with QuickPlot, you can be unlucky and define a range that happens to *force* Mathcad to evaluate, and therefore plot, a value of a function that is has very large value, swamping the vertical scale. Try plotting $w(x, y)$ from Fig. A.5 with $y = 3$ and an x range 0,001..4!

$$x := 0..20$$

$$f(x) := x \cdot \sin(x) \qquad g(x) := e^{-\frac{x}{5}} \cdot \cos(2 \cdot x) \cdot \Phi(x - 1)$$

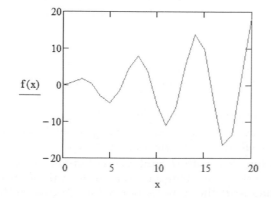
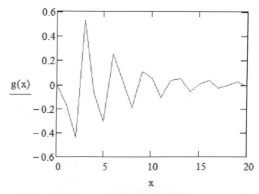

Figure A.16
An example of poor use of a range

$x := 0, 0.1 .. 20$

$f(x) := x \cdot \sin(x)$ $\qquad g(x) := e^{-\frac{x}{5}} \cdot \cos(2 \cdot x) \cdot \Phi(x - 1)$

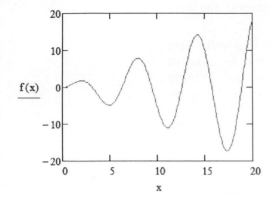

Figure A.17
An example of good
use of a range

2. Plotting more than one function on the vertical axis (But see immediately below this list for a big benefit of this approach over the QuickPlot approach!).

3. Using units.

4. Making parametric plots.

One advantage of this approach is that you directly control the horizontal scale of the graph with the range you define. An even better benefit is that *you can now plot several functions over differing ranges on the same graph*!

Figure A.18 demonstrates this. We plotted $f(x)$ from $x = 1$ to $x = 4$, and $g(x)$ from $x = 0$ to $x = 8$ by defining the two ranges shown. We can make some points here:

1. Regardless of the arguments with which functions are defined, you can plot using arguments of your choice. In Fig. A.18 we defined both f and g as functions of x, but plotted g as a function of x'. (To get the ´, type \`, not the prime key!)

2. You can plot up to 16 functions, with the following rules:

a. If *all* functions use the same argument (e.g., $f(x)$, $g(x)$, $h(x)$, . . .) you only need that one variable (e.g., x) on the horizontal axis.

b. If you wish to use functions with several different arguments (e.g., $f(x), g(x'), h(x''), \ldots$), you must match function arguments on the vertical axis with an equal number of the correctly ordered arguments on the horizontal axis (e.g., x, x', x'', . . .), with each argument range being separately defined. Actually, this is not always true, but you don't want to know about the exceptions!

$$f(x) := x \cdot \sin(x) \qquad g(x) := e^{-\frac{x}{5}} \cdot \cos(2 \cdot x) \cdot \Phi(x - 1)$$

$$x := 1, 1.01 .. 4 \qquad x' := 0, 0.01 .. 8$$

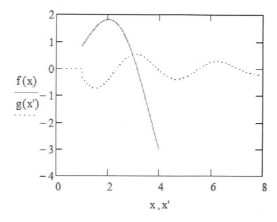

Figure A.18
Plotting functions with
differing ranges

Plotting One or More Arrays (Vectors)

Plotting arrays (actually, column arrays, or vectors) is similar to plotting functions in the previous section. Figure A.19 shows an example. We created three columns of numbers x, y, and z (but obviously the data could come from anywhere, e.g., it could be imported data). We then inserted a graph and added the arguments as shown. Note that y and z are not *functions* of x. We are just correlating, by plotting, sets of data; we could equally well have plotted y and x against z. As before, you can still set upper and lower limits on the vertical or horizontal axes.

To plot an array, the most important point is, *you must plot it against an array of the same size*. Often, when we plot data (i.e., arrays), we wish to present the results as a bar chart. To do this, start with an X-Y Plot, and then change the formatting (see the section on formatting below).

Plotting Arrays (Vectors) and Functions

Plotting arrays and functions together is fairly straightforward if you observe a few basic rules.

1. If the functions use the same argument as the array used for the horizontal axis, you only need to specify that array once on the horizontal axis. This is illustrated in Fig. A.20. Comparing Figs. A.19 and A.20, we see that in Fig. A.20 z is a *function* of x (in Fig. A.20 we evaluate z, but we didn't need to); hence we must use $z(x)$ on the vertical axis, not z!

$i := 0 .. 100$ $x_i := \dfrac{i}{10}$ $y_i := \sin\left(\dfrac{i}{10}\right)$ $z_i := e^{-\frac{i}{50}}$

	0
0	0
1	0.1
2	0.2
3	0.3
4	0.4
5	0.5
6	0.6
7	0.7
8	0.8
9	0.9
10	1
11	1.1
12	1.2
13	...

x =

	0
0	0
1	0.1
2	0.199
3	0.296
4	0.389
5	0.479
6	0.565
7	0.644
8	0.717
9	0.783
10	0.841
11	0.891
12	0.932
13	...

y =

	0
0	1
1	0.98
2	0.961
3	0.942
4	0.923
5	0.905
6	0.887
7	0.869
8	0.852
9	0.835
10	0.819
11	0.803
12	0.787
13	...

z =

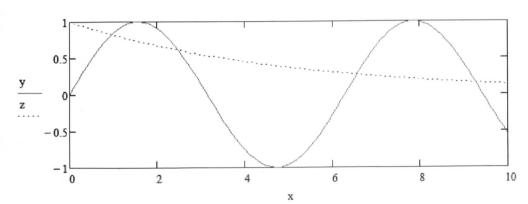

Figure A.19
An X-Y Plot of arrays

$$i := 0..100 \qquad x_i := \frac{i}{10} \qquad y_i := \sin\left(\frac{i}{10}\right) \qquad z(x) := e^{-\frac{x}{10}}$$

x =

	0
0	0
1	0.1
2	0.2
3	0.3
4	0.4
5	0.5
6	0.6
7	0.7
8	0.8
9	0.9
10	1
11	1.1
12	1.2
13	...

y =

	0
0	0
1	0.1
2	0.199
3	0.296
4	0.389
5	0.479
6	0.565
7	0.644
8	0.717
9	0.783
10	0.841
11	0.891
12	0.932
13	...

z(x) =

	0
0	1
1	0.99
2	0.98
3	0.97
4	0.961
5	0.951
6	0.942
7	0.932
8	0.923
9	0.914
10	0.905
11	0.896
12	0.887
13	...

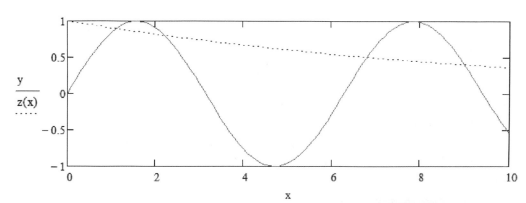

Figure A.20
An X-Y Plot of an array
and a function

$x' := 2, 2.01 .. 8$

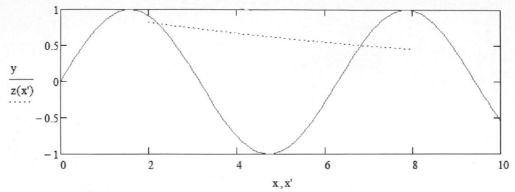

Figure A.21
Defining a range for
the function

2. If you wish to plot functions, along with arrays, with arguments other than the array on the horizontal axis, you must *define the range you wish to plot, and use that range on the horizontal axis on a one-for-one basis.* Figure A.21 shows an example of this, where we only plot the function $z(x')$ over a range $x' = 2$ to $x' = 8$.

As when we plotted functions and when we plotted arrays, you can set upper and lower limits on the vertical or horizontal axes.

Formatting X-Y Plots

Most of the graphs we've seen so far in this appendix are fairly crude. While it's not as good as Excel for formatting two-dimensional plots, Mathcad has quite a few bells and whistles for doing so. To access formatting options, simply double-click on a graph to get the Formatting Currently Selected X-Y Plot window shown in Fig. A.22*a*. This shows a number of options for changing formatting, most of which are obvious in their effect. Figures A.22*b* and A. 22*c* show other tabs for the same window.

Note that Fig. A.22*b* has more options you can scroll to; an important one is *Type*, which allows you to change from *line* to, for example, *bar* or *solidbar* for creating bar charts!

For an example of formatting, double-click on the graph in Fig. A.15; we can format it to look like Fig. A.23.

A.3

Polar Plots

You can create a Polar Plot by doing any of the following (in order of convenience):

1. Type **Ctrl + 7**.
2. Click the Polar Plot icon, ⊕, on the Graph Toolbar.
3. Use *Insert . . . Graph . . . Polar Plot.*

You will obtain a blank graph, as shown in Fig.A.24, with four placeholders; two are *required* (▪), and two are *optional* (▪).

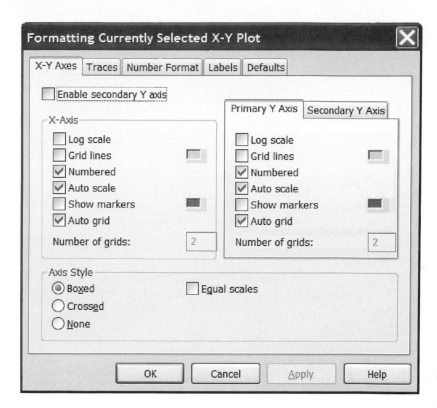

Figure A.22a
The window for X-Y
Plot formatting

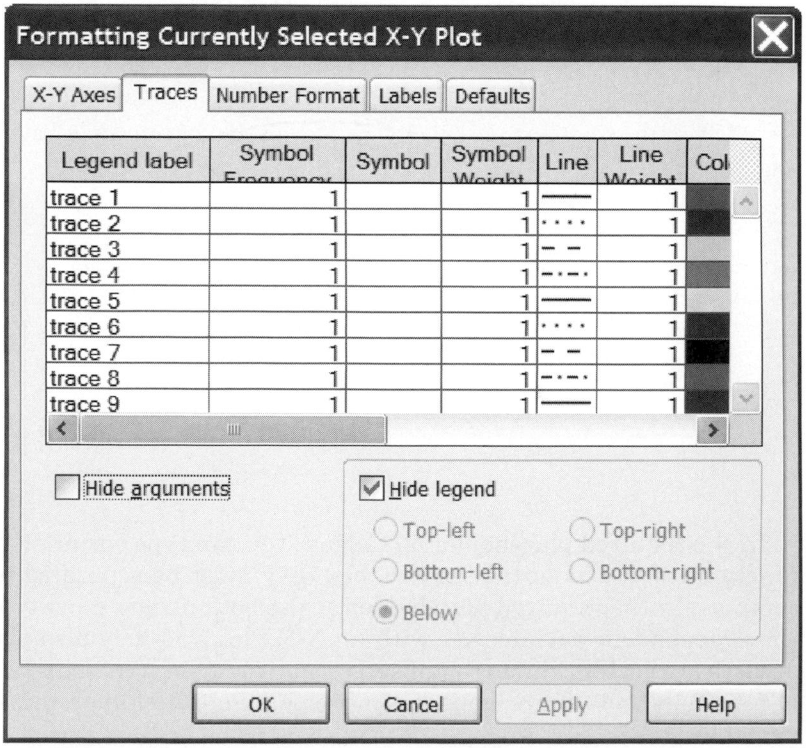

Figure A.22b
The Trace tab

Formatting Currently Selected X-Y Plot ✕

| X-Y Axes | Traces | Number Format | Labels | Defaults |

Title

[   ]

◉ <u>A</u>bove   ○ <u>B</u>elow   ☑ <u>S</u>how Title

Axis labels

☑ <u>X</u>-Axis: [   ]

☑ <u>Y</u>-Axis: [   ]

☑ <u>Y</u>2-Axis: [   ]

[OK]   [Cancel]   [<u>A</u>pply]   [Help]

***Figure A.22c
The Labels tab***

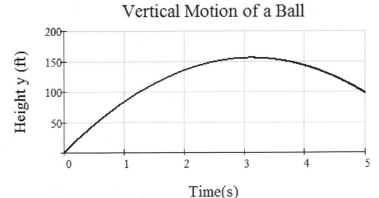

Vertical Motion of a Ball

***Figure A.23
Figure A.15 after
reformatting***

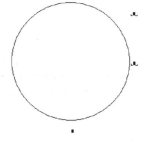

In the required placeholder on the left, you can type one or more functions of the independent variable; they must be separated by commas. In the required placeholder at the bottom, you must type the independent variable. As with an X-Y Plot, you may also plot an array (a vector column) against a second array, or a combination of functions and arrays against a corresponding set of appropriate variables.

***Figure A.24
A blank Polar Plot***

The optional placeholders are for inserting lower and upper limits on the radial scale; otherwise, Mathcad's QuickPlot feature automatically scales the graph based on function minima and maxima. The polar plot always presents a full 360° (or 2π) circle, and QuickPlot automatically plots functions over the complete circle, although you can define a range to plot.

Figure A.25 shows a plot of two functions; Fig. A.26 show the same functions plotted, with one plotted over a defined range.

For details on setting up and formatting a Polar Plot, review the sections entitled "Plotting One or More Functions using QuickPlot," "Plotting One or More Functions by Defining a Range or Ranges," and "Formatting X-Y Plots"; most of the information there applies here.

$$r(\theta) := 2 + \sin(3 \cdot \theta)$$

$$R(\theta) := 3 + \cos(7 \cdot \theta)$$

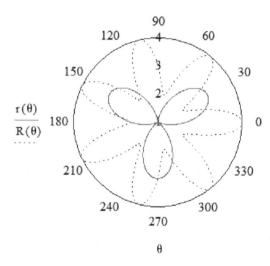

Figure A.25
A Polar Plot of two functions

You can create a Surface Plot by doing any of the following (in order of convenience):

1. Typing **Ctrl + 2.**
2. Clicking the Surface Plot icon, 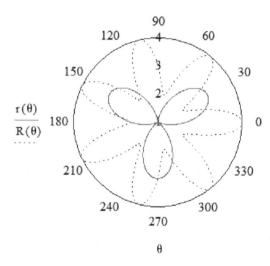, on the Graph Toolbar.
3. Using *Insert . . . Graph . . . Surface Plot.*

$$\theta' := 0, \frac{\pi}{100} .. \pi$$

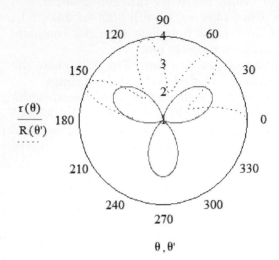

$$\frac{r(\theta)}{R(\theta')}$$

$$\theta, \theta'$$

**Figure A.26
A Polar Plot with a
defined range**

You will obtain a blank graph with a single placeholder (∎), in which you can type one or more variable names. There are several ways to generate a surface plot, including:

1. Defining a function of two variables. Figure A.27 provides an example; it shows two functions of two variables, $f(x, y)$ and $g(x, y)$. In the placeholder you type **f,g** (not **f(x,y)**, **g(x,y)**!); Mathcad uses QuickPlot to plot the two surfaces for x and y both varying from –5 to 5. Figure A.27 is particularly

$$f(x,y) := x^2 + y^2 \qquad g(x,y) := 20 \cdot \sin(x) \cdot \sin(y)$$

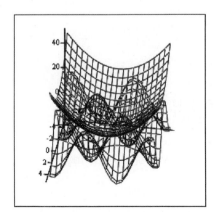

**Figure A.27
A Surface Plot of two
functions**   f ,g

ugly, but you can double-click on it to get the 3-D Plot Format window shown in Fig. A.28; there are an *amazing* number of bells and whistles here. Figure A.28 shows the tab for the QuickPlot data, but you should explore them all! Using this window, a little effort converts the graph in Fig. A.27 to that in Fig. A.29.

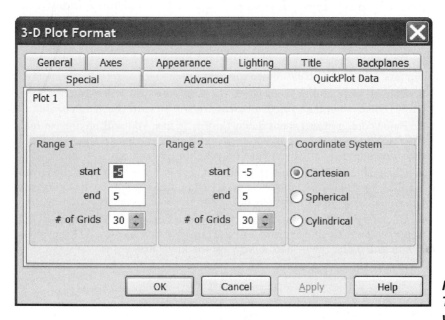

*Figure A.28
The 3-D Plot Format
window*

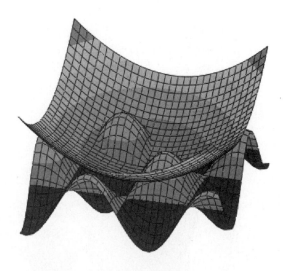

f , g

*Figure A.29
A formatted Surface
Plot*

2. Defining a matrix of values parametrically. Figure A.30 show an example of this.

3. Defining the x-, y-, and z-coordinates parametrically. Figure A.31 shows an example of this (with the same resulting graph as in method 2 above). Note that to make x, y, and z plot as coordinates (rather than three different functions), they *must* be enclosed in parentheses on the graph.

Note that you can also change the type of three-dimensional plot; Fig. A.32 shows the tab on the 3-D Plot Format window that you use to do this.

Here are some fun (and useful) little extras:

1. Click and click-hold (i.e., double-click but don't release the mouse) on a Surface Plot to real-time rotate the graph.

2. Before doing step 1, hold down the **Ctrl** key; you will real-time zoom (And I mean zoom!) as you drag away from or toward the graph center.

3. Before doing step 1, hold down the **Shift** key; you will be able to real-time rotate as before, but when you release the mouse the graph will continue spinning until you click on it again.

4. Click on an X-Y Plot and then click on the Zoom icon, ⊕, on the Graph Toolbar. You'll be able to zoom in on a region of the graph by dragging over the region and clicking the *OK* button on the zoom window.

5. Click on an X-Y Plot and then click on the Trace icon, ✹, on the Graph Toolbar. You'll be able to trace, or read out, graph values by clicking on the graph and using the ← and → keys on the keyboard.

$$F(u) := \begin{pmatrix} 5 - 2 \cdot u \\ u^2 \\ \sin(2 \cdot u) \end{pmatrix}$$

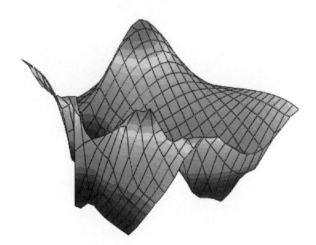

**Figure A.30
A Surface Plot of a
matrix of values**

$$x(u) := 5 - 2 \cdot u \qquad y(u) := u^2 \qquad z(u) := \sin(2 \cdot u)$$

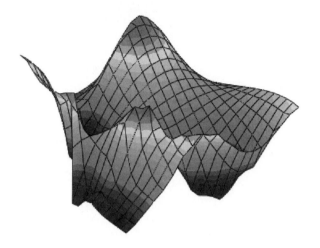

$$(x, y, z)$$

Figure A.31
A parametric Surface
Plot

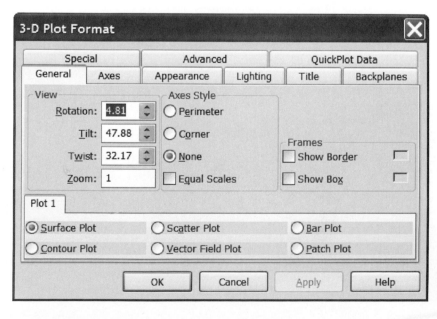

Figure A.32
Selecting a 3-D Plot
type

Answers to Selected Exercises

2.2 $V = 98.2$

2.4 $V = 32.7$

2.6 $A = 68.3$

2.8 $A = 3.49$

2.10 $x_c = 2.65$

2.12 $T = 330$

2.14 (a) 1   (b) 0.5   (c) 0.224   (d) 0.434   (e) 0.434

2.20 $f(0) = 5$ $f(1) = 0$ $f(5) = 0$

2.22 $f(0) = 1$ $f(1) = 0.988$ $f(5) = -0.756$

2.24 $f(0,0) = 1$ $f(1,1) = 0.442$ $f(1,5) = 0.232$ $f(5,1) = 0.199$
$f(5,5) = 0.104$

2.26 $f(0,0) = 1$ $f(1,1) = 0.905$ $f(1,5) = 0.273$ $f(5,1) = 0.273$
$f(5,5) = 0.082$

2.28 $V = 76.5$ gal $V = 290$ L $M = 237$ kg $M = 524$ lb

2.30 $P = 26.3$ hp

2.32 $E = 324$ calories (diet)

2.34 $I = 0.216$ $df/dt = 0.169$ $d^2f/dt^2 = 0.685$

2.36 $I = 1.852$ $df/dt = 0$ $d^2f/dt^2 = -0.333$

2.38 $x_{CG} = 16.8$ in $y_{CG} = 8.27$ in

2.40 $p_{systolic} = 2.27$ psi, 15.7 kPa $p_{diastolic} = 1.51$ psi, 10.4 kPa

2.42 $y(0.5s) = 40$ ft $y(1s) = 71.9$ ft $y(2s) = 112$ ft $v(0.5s) = 49$ mph
$v(1s) = 38.1$ mph $v(2s) = 16.1$ mph

2.44 $I(0s) = 4$ mA $I(1s) = 1.80$ mA $Q = 0.25$ J $E = 0.25$ J

3.2 $\theta = 10°$ $P = 35$ m

3.4 $t(100\text{ft}) = 1.61$ s, 3.86 s $t(120\text{ft}) = 2.59$ s, 2.88 s

3.6 (a) $f = 0.0502$ $\Delta p = 3.56$ psi   (b) $f = 0.0417$ $\Delta p = 47.4$ psi

3.8 $x = 5.465$

3.10 $x = 0.375, 1.571, 2.767, 4.712$

3.12 $x = -2, -1, 1, 3$

3.14 $x = -2, -1, 1, 3$

3.16 $I_1 = -0.48\,\text{A}$ $I_2 = -0.96\,\text{A}$ $I_3 = 0.6\,\text{A}$ $Q = 15.1\,\text{W}$

3.18 $x = -1.287, y = -0.574$ $x = 2.487, y = 6.974$

3.20 $x = 0.0686, y = 1.0686$ $x = 0.877, y = 1.877$
   $x = 2.223, y = 3.223$ $x = 2.732, y = 3.732$

3.22 $r = 24.6\,\text{m}$ $\theta = 35.3°$

3.24 $r = 15\,\text{in}$ $h = 3.65\,\text{in}$

3.26 $X = 13.2\,\text{ft}$

3.28 $R = 0.573\,\text{cm}$

3.30 $P = 40,000\,\text{lbf}$ $P = 30,000\,\text{lbf}$

3.32 $p = 170\,\text{kPa}, 24.6\,\text{psi}$

3.34 $\omega = 31.8\,\text{Hza}$

3.36 $x = 23.6\,\text{cm}$

3.38 $x = 25.2\,\text{cm}$

3.40 $x = -2.678$ $x = 0.464$ $x = 3.605$

3.42 (a) $h = 40\,\text{cm}$ (b) $h = 101\,\text{cm}$

3.44 $I_1 = -0.48\,\text{A}$ $I_2 = -0.96\,\text{A}$ $I_3 = 0\,\text{A}$ $I_4 = -2.4\,\text{A}$

3.46 $x = 0.783$ $x = 26.36$ $x = 39.89$

3.48 $Q = 4.75\,\text{gal/min}$

Chapter 4

4.2 $3\cdot\text{in}, 6\cdot\text{in} .. 2\cdot\text{ft}$

4.4 $27, 22 .. 2$

4.6 $\max(M) = 0.745$ $\min(M) = -0.745$

4.8 2.717

4.12 $\max(V) = 0.22\,\text{m/s}$

4.14 $|\mathbf{d}| = 13.4$ Angles $= 14.1°, 23.7°, 36.1°$

4.16 $(0.391, 0.13, -0.911)$

4.18 Work $= 10\,\text{J}, 7.38\,\text{lbf}\cdot\text{ft}$

4.24 $x = \begin{bmatrix} 0.375 & 2 & 1.5 & 1 \\ -0.438 & 1 & 0.75 & 0 \\ 0.375 & -1 & -0.5 & 0 \end{bmatrix}$

4.26 $x = \begin{bmatrix} 0 & 1 & 0 \\ 2 & -3 & -3 \\ 3 & -3 & 0 \\ 0 & 1 & -3 \end{bmatrix}$

4.28 Black $= \$32$ Red $= \$20$ Green $= \$24$ White $= \$16$

4.30 $I_1 = -0.48\,\text{A}$ $I_2 = -0.96\,\text{A}$ $I_3 = 0.6\,\text{A}$ $Q = 15.1\,\text{W}$

4.32 $I_1 = 0.24\,\text{A}$ $I_2 = -0.72\,\text{A}$ $I_3 = 0\,\text{A}$ $I_4 = -2.4\,\text{A}$ $Q = 67.7\,\text{W}$

4.34 $I_1 = 0.535$ A $I_2 = -0.129$ A $I_3 = -0.591$ A $I_4 = -2.69$ A
$Q = 110$ W

4.36 $f = 4.94$ Hz $f = 8.84$ Hz

4.38 $f = 5.11$ Hz (same frequency as $m = 15$ lb, $k = 40$ lbf/in) $f = 1713$ Hz

Chapter 5

5.2 (a) $\cos(\theta)^2 - \sin(\theta)^2$ (b) $2 \cdot \cos(\theta) \cdot \sin(\theta)$

5.4 $\dfrac{3}{x-3} - \dfrac{4}{x+2} + \dfrac{2}{x+4}$

5.8 e^x

5.10 $\sin(\theta)$

5.12 $\gamma (= 0.57721566490153286)$

5.14 $a \cdot c - b \cdot d + (a \cdot d + b \cdot c) \cdot i$

5.16 $\dfrac{1}{\theta^2 + 1}$

5.18 $t = -\tau \ln\left(\dfrac{T}{T_0}\right)$

5.22 $\theta = \pi/4$

5.32 $\cos(2\pi x) \approx 1 - 2\pi^2 x^2 + \dfrac{2\pi^4 x^4}{3}$

$\cos(2\pi x) \approx 1 - 2\pi^2 x^2 + \dfrac{2\pi^4 x^4}{3} - \dfrac{4\pi^6 x^6}{45}$

$\cos(2\pi x) \approx 1 - 2\pi^2 x^2 + \dfrac{2\pi^4 x^4}{3} - \dfrac{4\pi^6 x^6}{45} + \dfrac{2\pi^8 x^8}{315}$

Chapter 6

6.4 $t = 6.24$ s $v_{max} = 97.5$ mph

6.10 $x(10) = -0.0291$ $y(10) = -0.176$

6.12 $B_{max} = 13.3$ mole/liter $t = 11.5$ hr

6.14 $\theta'(0) = 2.53$ rad/s, 3.93 rad/s, 5.43 rad/s, 6.95 rad/s

Chapter 7

7.2 $W_{max} = 2252$ $W_{min} = 1454$ $W_{mean} = 1944$
(a) 2   (b) 3   (c) 6

7.4 slope = 1.34   intercept = -1.84 corr = 0.996
(a) $y(2.5) = 1.51$ (b) $y(5.5) = 5.53$ (c) $y(11) = 12.9$
(d) $x(y = 5) = 5.10$ (e) $x(y = 13) = 11.1$

7.6 slope = 1.93   intercept = 2.78   corr = 0.995   (a) $y(1.5) = 5.67$
(b) $y(8.5) = 19.2$ (c) $y(-1) = 0.851$ (d) $x(y = 10) = 3.74$
(e) $x(y = 24) = 11.0$

7.8 Correlation = 0.978 (good)   Sales($35,000) = 7560   Sales($0) = 2530

7.10 $a = -2.24$ $b = 0.28$ $c = 2.46$ $y(6) = -9.53$ $x(y= 1.5) = -3.03$

7.12 $a = 1.91$ $b = -0.271$ $y(12) = 4.49$ $x(y= 4.5) = 12.1$

7.14 $a = 0.662$ $b = 1.49$ $c = 0.632$ $y(5.5) = 9.00$ $x(y= 5) = 3.55$

7.16 $a = 1.49 \times 10^{-3}$ $b = 2.71$ $c = 6.72$ $P(75\text{mph}) = 183$ hp
$V(55 \text{ hp}) = 46.4$ mph

7.18 $a = 2.66$ $b = 1.60$ $c = 0.672$ $y(10) = 2.65$ $x(y= 2.5) = 4.83$

7.20 $a = 13.6$ $b = 1.57$ $c = -0.855$ $y(-1) = 11.2$ $x(y= 4.5) = 4.32$

7.22 $a_0 = 1.36$ $a_1 = 1.76$ $C(1.5) = 2.18$ $t(C= 2\ \mu\text{m}) = 1.24$

Index